Computer-Aided Kinetics
for Machine Design

MECHANICAL ENGINEERING

A Series of Textbooks and Reference Books

EDITORS

L. L. FAULKNER

Department of Mechanical Engineering
The Ohio State University
Columbus, Ohio

S. B. MENKES

Department of Mechanical Engineering
The City College of the
City University of New York
New York, New York

1. Spring Designer's Handbook, *by Harold Carlson*
2. Computer-Aided Graphics and Design, *by Daniel L. Ryan*
3. Lubrication Fundamentals, *by J. George Wills*
4. Solar Engineering for Domestic Buildings, *by William A. Himmelman*
5. Applied Engineering Mechanics: Statics and Dynamics, *by G. Boothroyd and C. Poli*
6. Centrifugal Pump Clinic, *by Igor J. Karassik*
7. Computer-Aided Kinetics for Machine Design, *by Daniel L. Ryan*

OTHER VOLUMES IN PREPARATION

Computer-Aided Kinetics for Machine Design

DANIEL L. RYAN

Associate Professor of Engineering Graphics
Clemson University
Clemson, South Carolina

MARCEL DEKKER, INC.
New York and Basel

Library of Congress Cataloging in Publication Data

Ryan, Daniel L., [date]
    Computer-aided kinetics for machine design.

    (Mechanical engineering; 7)
    Includes bibliographies and index.
    1. Machinery--Design--Data processing. 2. Machinery,
Dynamics of--Data processing. I. Title. II. Series.
TJ233.R85   621.8'15'02854   81-3146
ISBN 0-8247-1421-0   AACR2

MARCEL DEKKER, INC.

270 Madison Avenue, New York, New York 10016

Current printing (last digit):
10 9 8 7 6 5 4 3 2 1

PRINTED IN THE UNITED STATES OF AMERICA

Preface

This book is the second in a computer-aided series which is currently
being published by Marcel Dekker, Inc. The first title, Computer-Aided
Graphics and Design, by the same author, was based on a series of lecture
notes that were used to present continuing engineering education seminars
on computer-aided graphics, design, and manufacturing. This second
title, Computer-Aided Kinetics for Machine Design, is an outgrowth of the
computer graphics course development study conducted at Clemson University during tne period 1977–1980. The resulting courses of study were
(1) EG 310 Computer-Aided Graphics, (2) EG 410 Computer-Aided Design,
and (3) EG 411 Computer-Aided Manufacturing Concepts.

Computer-aided design (CAD) and its impact on manufacturing is one
of the revolutionary concepts of the twentieth century. CAD presents not
only engineering, but also management and production problems, whose
impact in one way or another will eventually affect all of society. Computer-aided graphics plays an important role in this field, for it is from
the computer-aided graphics database that the initial information is obtained for the designer and for the eventual production of a machine part by
a computer-controlled machine tool.

This book is intended to acquaint the reader with the computer-aided
design of machine elements and the relation of computer-aided graphics to
it. The reader should be familiar with the concepts of computer graphics,
machine design, and computer programming before attempting any of the
examples that vividly show this relationship. The selection of material in
this text is based on the premise that the reader of the first five chapters
has had a course in kinematics, while a reader of the second five chapters
has had exposure to the design of machine elements. Therefore, many
basic design situations are included without lengthy explanations or manual
solutions. This makes it possible to keep the emphasis on computer-aided
design graphics.

Methods of using existing computer-aided machine design programs are stressed, as is the procedure for writing new ones. The discussion, however, is on the use rather than the creation. It is the author's belief that this delimitation is necessary in a computer-aided machine design course, because computer-aided design requires more ingenuity, inventiveness, imagination, and patience than display graphics. An ability to create new machine designs will be developed by experience after the student (user) learns the basic computer application techniques described in this book. Also, a complete bibliography is included at the end of each chapter for those who wish to delve further into this concept. The unique features of this book include the following:

1. It is a study of computer-aided machine design—not a computer text or a designers handbook of cookbook solutions.
2. Fundamental concepts of kinematics and machine element design are explained in lay terms as they relate to computer displays.
3. The book has a common illustrative pattern for showing the CRT computer-generated examples produced from a computer program.
4. A unique combination of computer-aided graphics and design skills are used in most of the problem solutions.

The author makes no claim to originality of the machine elements used as examples. This book traces the history of certain machine elements and their uses; therefore this book draws heavily on the earlier works of others. It is a pioneer effort in the computerization techniques of some heretofore manual methods, however. The references consulted are included in the chapter bibliographies and in various footnotes. It is the author's hope that those concerned with developing new programs in CAD, in both industry and education, will find this book a source of useful information. The author is grateful to the industrial manufacturers and users of computer-aided design equipment for the many hours of conferences and visits, and for supplying considerable information on the many practical uses of computer-aided design. Every CAD example supplied by a manufacturer has been identified by a courtesy line.

The author deeply appreciates the kindness and generosity of those persons, some students and some teachers, who have made valuable contributions to this reference text. Most of all, special consideration was shown the author during the writing of this second text in computer-aided applications from Connie, Tim, and Tess. The author's indebtedness to an understanding family is hereby reaffirmed.

Daniel L. Ryan

Contents

Preface iii

1. Introduction 1

Introduction to Computer-Aided Machine Design 1
Computer Display of Machine Elements 3
Computer Analysis of Machine Elements 6
Summary 17
Appendix—A Brief History of Computer-Aided
 Machine Design 19
Bibliography 23

2. Computer Display of Machine Motion 25

Animation of Machine Parts 25
Simulation of Machine Part Motion 33
Documentation of Machine Motion as Computer Output 36
Summary 38
Appendix—TSO Commands 40
Bibliography 43

3. Computer-Aided Velocity Analysis for CRT Displays 45

Velocities in Computer-Aided Machine Design 45
Centros in Computer-Aided Analysis of
 Machine Elements 65
Summary 68
Bibliography 70

4.   Interactive Acceleration Analysis   71

     Acceleration in Machine Design   71
     Summary   85
     Appendix—Graphical Interactive Programming (GRIP)   85
     Bibliography   89

5.   Computer-Aided Linkage Design   91

     Four-Bar Linkage Displays   91
     Four-Bar Linkage Situations   99
     Linkage and Sliding Members   106
     Summary   117
     Bibliography   118

6.   Computer Generation of Transmission Paths   121

     Transmission of Motion by Computer Modeling   121
     Plotting the Conjugate to a Given Path   131
     Display of a Curve in Rolling Contact   133
     Machine Elements in Rolling Contact   137
     Summary   140
     Bibliography   141

7.   Gears and Cams   143

     Types of Gear Displays   143
     Gear Elements and Display Techniques   146
     Gear Studies and Computer Analysis   153
     Types of Cam Displays   157
     Design Programs   162
     Summary   172
     Bibliography   172

8.   Computer-Aided Design of Flexible Connectors   175

     Computer Solution of Pitch Surfaces   175
     Computer Display of Drivers and Machine Elements   182
     Inventory and Selection by Computer Program   192
     Summary   207
     Bibliography   208

9. Computer-Matched Machine Elements   209

   Computer-Selected Trains   209
   Display of Example Trains   216
   Design of Gear Trains   218
   Summary   221
   Bibliography   221

10. Computer-Aided Combination of Machine Elements
    in the Design of Working Machinery   223

    Aggregate Computer-Aided Combinations   235
    Chapter and Book Summary   238

Tests   243

    1. Introduction to Computer-Aided Design   245
    2. Computer Display of Machine Motion   249
    3. Computer-Aided Velocity Analysis for CRT Displays   253
    4. Interactive Acceleration Analysis   255
    5. Computer-Aided Linkage Design   257
    6. Computer-Generated Transmission Paths   259
    7. Gears and Cams   261
    8. Computer-Aided Design of Flexible Connectors   263
    9. Computer-Matched Machine Elements   265
    10. Computer-Aided Combination of Machine Elements
        in the Design of Working Machinery   267

Index   269

Computer-Aided Kinetics for Machine Design

1

Introduction

Computer-aided design (CAD), now central to an engineering education, is also used in many other professions and specialties: medicine, architecture, pure sciences, arts and humanities, and so forth. In industry CAD is used for products, apparatus, machines, systems, tools, structures, circuits, and in many other applications. CAD is a dynamic, new field of study and includes several subtypes. In the case of computer-aided machine design (CAMD), the specialization covers everything from a ball bearing 1mm in diameter to a complete manufacturing plant. It may be used to design a small piece part or a jet engine for an aircraft. The user of a CAMD system performs analytical investigations and tests and studies various computer programs for modeling such things as environmental effects. A computer-generated design requires answers to at least the following questions:

1. What device, mechanism, or machine element should be used in the final machine?
2. What system components should be used: mechanical, electrical, pneumatic, hydraulic, or other?
3. What material selection is available?
4. What are the loads on the machine members?
5. How will the size, space, and weight of machine elements affect the cost of the design?

INTRODUCTION TO COMPUTER-AIDED MACHINE DESIGN

When designing a machine, or studying the design of an existing machine, two closely related parts of the procedure present themselves. First, the machine parts must be so proportioned and related to each other as to provide proper motion. Second, each part must be adapted to withstand the forces imposed upon it. The nature of the part movement does not

depend upon the strength or absolute dimensions of the moving parts.
This can be shown by computer models whose dimensions are variable
and may differ from those required for strength, yet the motions of the
model will be the same as those of the machine part.

The force and the motion may be modeled separately, dividing
computer-aided machine design into two sciences:

1. <u>Kinematics of machines</u>, which is a computer study of the motion
    and forms of the elements of a machine, and the method of sup-
    porting or guiding each element separately from strength
    considerations.
2. <u>Machine mechanism design</u>, which involves the calculation and
    display of the forces acting on different elements of the machine.
    The selection of materials, based on strength and other
    characteristics, to withstand the applied forces can also be
    modeled and displayed by the computer.

There are a number of ways of approaching a machine design problem
for which the computer can be used as a solution tool. There are no set
rules for the designer to follow. A commonsense approach in the use of
the computer for the design of a new machine may be used.

1. Make a complete statement of the purpose for which the
    machine is to be designed — the computer won't tell you!
2. Review current computer files and select the library of
    mechanisms that could give the desired motion or group
    of motions.
3. Display the mechanisms on an output device such as a cathode
    ray tube (CRT) to study the forces acting on each member of
    the mechanism selected.
4. Select a mechanism and use it in a material comparison program
    to assist in the selection of the material best suited for each
    member.
5. Size each member by running a program designed to consider the
    forces, permissible stress for the selected material, and the
    deflection or deformation.
6. Modify the members of the mechanism by the interactive computer
    method so that previous design experience facilitates manufacture.
7. Use the design database collected in steps 1 through 6 to output
    assembly and detail drawings of the machine, including material
    specifications and computer-aided manufacturing instructions.

Often, situations arise in which size depends upon space limitations. Many
members must be a certain size to fit the application or cause the desired
motion. A mechanical engineer will use standard sizes whenever possible,

changing the computer size to a less costly standard-sized part. A computer
science-oriented designer is often prone to rely entirely upon methematical
analysis, but in certain situations an exact computer analysis of stresses
and deformations cannot be made. The designer then resorts to his experience to proportion and manufacture the part properly.

COMPUTER DISPLAY OF MACHINE ELEMENTS

In order to display machine elements by computer assistance, the designer
must be familiar with the techniques of computer graphics. A designer
thinks in symbols and pictures, whereas the computer understands only
simple electrical impulses. The solution is to have the person and the
machine work together as a team. Over the past decade, computer
graphics displays, particularly for CAD, have been justified because they
can save time and money. Figure 1.1 illustrates the display configuration
necessary for CAMD.

The computer graphics display is a way of converting the computer's
impulses into engineering documents and, conversely, to translate the
designer's instructions (in picture form) into electronic data. In many
of the more sophisticated CAD systems, the designer doesn't need to know
about electronics or computer construction in order to control the CAMD
function. In general, computer displays for machine elements include any
device that converts computer language to people language, or any device
that converts people language to computer language, with the intent of
solving problems by creating graphical images.

A computer display is composed of line images, notation, and special
symbols used by the designer. It is a representation of a combination of
members arranged such that they represent a machine. A designer
represents a machine as an assemblage of parts displayed between the
power source and the work. Each member in a machine design either
moves or helps to guide the motion of some of the other components.

As shown in Figure 1.1, the computer display can be either a CRT
presentation or a plotter drawing. For quick copies the hard-copy device
can copy the contents of the CRT screen. The graphics tablet is not a
display device; rather, it is used to input graphical information to the
CAMD system. The designer may view the input on the CRT before
placing it in computer storage.

Cranks, levers, bearings, journals, shafts, cams, gears, threads,
pulleys, stops, cylinders, rods, chains, links, relays, trips, pins, keys,
collars, and plates are common forms of the parts that can be displayed
by the computer. Many of these machine parts in this book will be presented as computer illustrations. A computer displays a CAMD symbol
by referring to a list of instructions written for the intended symbol. In
the case of the point on a member, the following set of instructions is

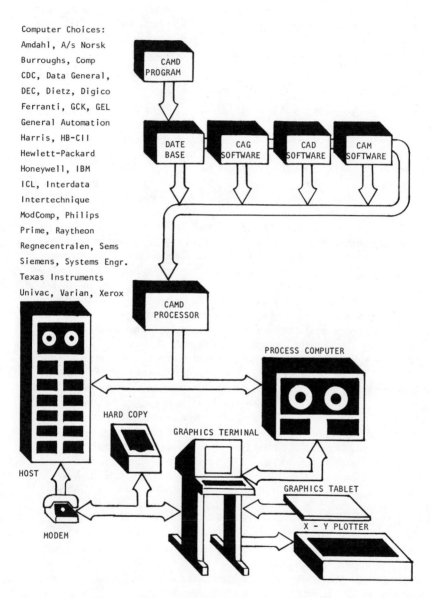

Computer Choices:
Amdahl, A/s Norsk
Burroughs, Comp
CDC, Data General,
DEC, Dietz, Digico
Ferranti, GCK, GEL
General Automation
Harris, HB-CII
Hewlett-Packard
Honeywell, IBM
ICL, Interdata
Intertechnique
ModComp, Philips
Prime, Raytheon
Regnecentralen, Sems
Siemens, Systems Engr.
Texas Instruments
Univac, Varian, Xerox

Figure 1.1 CAMD display configuration.

```
      SUBROUTINE DOT(XRB, YRB, DIARB)
      XIRB=XRB+. 5*DIARB
      RADRB=DIARB*. 5
      CALL CIRC L(XIRB, YRB, 0. , 3600. , RADRB, 0. , 0. )
      RETURN
      END
```

The designer may place the "dot" anywhere on the display by entering the command

      CALL DOT(XRB,YRB,DIARB)

where XRB and YRB are the coordinate locations of the center of the dot and DIARB is the diameter of the dot to be displayed。 Each item in Table 1.1 has a separate list of instructions.

Each of the calling sequences must begin with the command

      CALL

for FORTRAN-based storage routines, followed by the name of the character set selected from Table 1.2. If BASIC routines are used, the command would be

      100 GOSUB 1000

where the number 100 represents a line number in the program and GOSUB is a command to go to the subroutine beginning on line 1000 in the BASIC program.

Table 1.1   CAMD Notation

Display character	Design interpretation
a	Angular acceleration
A	Linear acceleration
●	Point on a member
T	Time
Θ	Fixed axis
V	Linear velocity
o	Pin joint
θ	Angular displacement
w	Angular velocity

Table 1.2   Notation Commands

Call sequence	Character	Design interpretation
a(X,Y)	a	Center of display
A(X,Y)	A	Lower left corner
DOT(XRB,YRB,DIARB)	o	Center of display
T(X,Y)	T	Lower left corner
FA(X,Y)	Θ	Center of display
V(X,Y)	V	Lower left corner
PJ(X,Y)	o	Center of display
THETA(X,Y)	θ	Center of display
OMEGA(X,Y)	w	Center of display

For the purpose of this discussion, think of each element as being graphical in nature. For a machine designer the emphasis is on the proper recall of the notation, not the particular language used to store each element in computer (host) storage.

COMPUTER ANALYSIS OF MACHINE ELEMENTS

A computer analysis for machine structures includes data on strength, stiffness, ductility, shock, fatigue resistance, creep, hardness, machinability, and other behavioral characteristics of the materials to be selected. By storing information in large computer files and updating it on a periodic basis, current data are available for all new materials and available fabrication methods.

Using the computer, the designer can display various tables of the most important engineering properties of materials commonly used in machine design; but the computer does not select the materials, the designer does. The designer must keep in mind the fabrication methods available and the effects of each on the properties of the structure member. The properties displayed in Table 1.3 vary with the mechanical and thermal treatment received during manufacture, their treatment during the structure usage, the size of the member, and the ambient conditions surrounding the machine structure. The values listed must be considered as averages only. As stated, higher values may be obtained by careful manufacture; lower values may result from certain internal and external conditions. The experienced machine designer

Table 1.3   Computer Display of Machine Materials

Material	Modulus E	Modulus G	Specific weight	Thermal expansion
Aluminum	10	3.8	0.095	013
Beryllium	44	—	0.065	006
Brass	17	5.3	0.309	010
Bronze	16	5.9	0.294	010
Cast iron	12	7.2	0.256	006
Copper	17	5.8	0.322	009
Lead	2.6	—	0.411	016
Magnesium	6.5	2.5	0.062	014
Nickel	30	—	0.310	007
Steel	30	11.5	0.282	006
Titanium	16.5	—	0.164	004

displays the computer file for reference because it is extremely fast
and accurate (up to date), then uses engineering judgment with regard
to the material under consideration.

Structures

A structure is a fabrication of elements capable of transmitting forces or
carrying loads but having no relative motion themselves. The designer
must select the proper materials for each structure member. This involves consideration of weight, size, and shape as well as the loads the
member must carry; the cost of the material and fabrication; and the
properties of the material used for the structure. An example of a
structure is the frame of a machine, consisting of several parts welded,
bolted, or fabricated to prevent motion between structure members.

Mechanisms

A mechanism is a combination of rigid bodies arranged such that the
motion of one compels the motion of the others. The computer-aided
design terms mechanism and machine are often used synonymously,
but correctly the combination is a mechanism when it is used to transmit or modify motion (a machine is used to transfer energy or perform
work). The combination of a crank, shaft, and connecting rod inside
the machine structure called a gasoline engine is a mechanism, because

reciprocating motion is converted into circular motion. For this example mechanism to become a useful machine, other mechanisms, such
valves, gears, and accessories, are added so that the energy of the
gasoline may be converted into work. Therefore, in CAMD, a machine
is a series of mechanisms, but a single mechanism is not necessarily a
machine.

Frames

The frame of a machine is a structure that supports the mechanisms (moving
parts, path regulation, and types of motion produced). Frames may be
either fixed in place or moving relative to the earth, as in the example
given above of the automobile engine. To analyze frames, it is more
convenient to use CAMD displays in determining, for example, the
endurance limits, ultimate strength, and yield stress. Because frames
support mechanisms, it is often necessary to consider the following:

1. Particles, the smallest part of a computer display. A particle
   represents a single point and was referred to earlier as a "dot."
   A line on a computer display may be thought of as a series of
   contiguous particles arranged side to side.
2. Rigid bodies, in which display particles remain at a constant
   distance from each other. In a computer display of a rigid
   member, the body is assumed not to suffer distortion by the
   forces acting on it.
3. A driver and a follower, parts of a mechanism that cause motion
   (driver) and whose motion is effected (follower).

For the purpose of CAMD analysis a line may be considered as being of
indefinite length and a body of indefinite magnitude. For example, in
analyzing the motion of a driver or follower, it may be necessary to
consider the motion of a point which is a part of that member but beyond
the limits of the actual body. The extended point must have the same
properties as all other points on the same member.

Modes of Transmission

For the purpose of computer display, the actions of natural forces for
attraction or repulsion are not programmed. One section of a mechanism
can not move another unless the two are in direct contact or are connected
together by an intermediate method that is capable of transferring the
motion. Therefore, motion can be transmitted from driver to follower
by direct contact (sliding or rolling) or by connectors, which may be
rigid, flexible, or fluid. If a connector is rigid, it is called a link.
Pivots called joints are necessary to connect the link to the driver and
follower. If the connection is flexible, a band is used and is capable
only of pulling a member. Fluids confined to a fixed space are also
excellent connectors.

Pairs of Elements

In order for a computer-aided machine element to be displayed as a moving body continually in contact with another body and at the same time moving in a definite path; a technique known as display pairs is used. This technique is extremely useful for plotter output or display on a direct-view storage tube (DVST; Figure 1.1).

The moving body shown in Figure 1.2 as element A will remain in contact with the other element, C, and at the same time move in a set path, B. The design of element A should have a shape, which can be found by allowing element A to occupy a series of positions related to element C and then displaying the series of positions. The program for displaying Figure 1.2 is as follows:

```
CALL  BEGIN
CALL  HPAIR(512,350)
CALL  FINITT(0, 0)
STOP
END
```

The subprogram for displaying HPAIR is as follows:

```
SUBROUTINE  HPAIR(IX,IY)
X=IX
Y=IY
CALL  CIRCL(X+2.,Y,0.,-90.,2.,2.,0.)
CALL  CIRCL(X+1.5,Y,0.,-90.,1.5,1.5,0.)
CALL  CIRCL(X+1.75,Y,90.,450.,.187,.187,0.)
CALL  CIRCL(X+1.5,Y-.5,90.,450.,.187,.187,0.)
CALL  CIRCL(X+1.25,Y-1.,90.,450.,.187,.187,0.)
CALL  CIRCL(X+.75,Y-1.25,90.,450.,.187,.187,0.)
CALL  CIRCL(X+.375,Y-1.375,90.,450.,.187,.187,0.)
RETURN
END
```

Therefore, if element A were displayed as shown in Figure 1.2, the form of C would be that of a circular channel. In order to design element A to move in a definite path B, A must be paired with C. The display is determined by the nature of the desired motion of the paired elements. The display program is rather short, depending upon the number and complexity of the pairs of elements. Subprograms are usually written to handle all standard elements used in the mechanical engineering design area. These subprograms are stored in the host computer shown in Figure 1.1. This type of computer display is called template storage (passive graphics) and is not recommended for real-time graphic displays in an interactive mode.*

*A complete description of template storage can be found in Ryan (1979, Chap. 4).

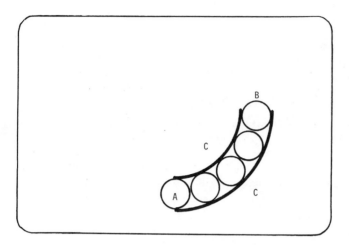

Figure 1.2 Computer display of pairs.

Lower Case.   When one element not only forms the envelope of the other
but also encloses it, a lower case exits. In lower cases, surface contact
exists between the two elements, one being full or solid, the other being
hollow or open. In machine design useful types of lower-case pairs can
be displayed by the computer and studied by the designer. They are:

1. A straight lined pair, allowing linear translation (ball and pipe)
2. A circle, allowing rotation (shaft and pillow block)
3. A helix, allowing a combination of straight translation and
   rotation (threaded shaft)

Higher Case.   The pair shown in Figure 1.2 is not a lower case, because
the element A can be removed from the open channel C. Ball and roller
bearings are examples of higher cases. These cases are easier to display
because sectional views are not required. Additional computer memory is
required to display lower cases because of the "closed" nature of the pair
of elements.

Incomplete.   Sometimes it is useful to study forces having a certain
definite direction that may affect the motion of the pair. When such
cases for computer display exist, it may no longer be necessary to
make the pair closed or higher. One element can be sectioned and cut
away where it is not needed to resist or balance forces. Examples are
thrust bearings, the ways of heavy machine tools, or other cases where
large loads and gravity form the incomplete pair.

<u>Inversion.</u> The relative motion of machine elements can be studied when both elements of the display pair move; however, one element of the pair is usually a fixed piece. In the case of a bolt and nut through a slot, the slot is usually fixed while the bolt and nut are free to move. In the case where the bolt is welded in place and a movable plate containing a slot is moved; an inversion of pairs occurs on the display screen. This exchange of the fixedness of an element of a pair with its partner, known as <u>inversion,</u> generally does not affect the absolute or relative motion of the lower- or higher-order pair.

To display this concept, output devices arranged as shown in Figure 1.3 are used. The pairs of elements are displayed on the face of the CRT. The display controller allows the machine designer to input new data from the 30 X 40 graphics tablet. The pair of elements can be modified and the new database, which now describes the inversion, can be stored in the 64 K memory module shown on the left. A hard-copy print of the modification may be produced by scanning the display CRT screen. Below the memory bus and multiplexer bus the normal CAMD hardware arrangement shown in Figure 1.1 is in operation. By this configuration a design from the lower portion can be sent to the display portion above for analysis and/or modification. If modifications are necessary, they are returned to the lower portion for permanent part storage on the host computer.

Bearings

The computer display <u>bearing</u> is, in general, that of the surfaces of contact between two machine elements that have relative motion. One element may support the other, one may be stationary, or both elements may be moving. These display programs are stored in computer memory according to the relative motions they will allow:

1. For straight translation, the bearing display must have plane or cylindrical surfaces, cylindrical being programmed as part of the software in most cases. If one element is fixed, the surfaces of the moving elements are called <u>slides;</u> the surface of the fixed element is called a <u>guide</u>.
2. For rotation, or turning, the display must have surfaces of circular cylinders, cones, conoids, or flat disks. The surface of the solid or full element is annotated on the CRT as a <u>journal,</u> <u>neck,</u> <u>spindle,</u> or <u>pivot;</u> the open or hollow element is labeled a <u>bearing, pedestal, pillow block,</u> or step.
3. For translation and rotation displays (helical motion displays), the elements must have a screw-thread shape. The full element will be labeled a <u>screw</u> and the open element a <u>nut</u>.

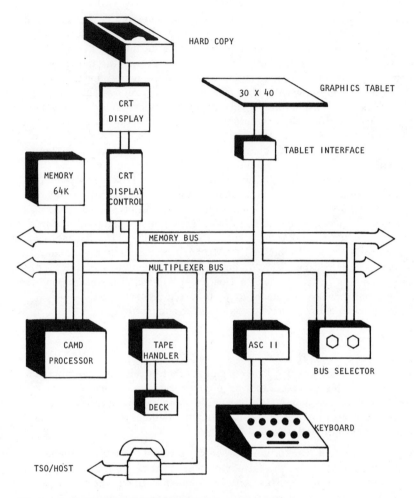

Figure 1.3 Expanded CAMD display configuration.

Collars and Keys

Often a computer-aided display of pulleys or wheels that turn freely on cylindrical shafts but have no longitudinal motion is useful. The CAMD database must provide rings or collars for this purpose. The subprogram COLLAR was used to display Figure 1.4. Collars D and E, held by set screws, prevent the motion of the pulley along the shaft but allow it free rotation.

When it is desirable to display no relative motion, axially or in circumference, the pulley or coupling may be fitted to the shaft by the

```
SUBROUTINE COLLAR(XPAGE, YPAGE)
CALL RECT(XPAGE, YPAGE, . 5, . 25, 0. , 3)
CALL RECT(XPAGE+. 25, YPAGE-. 25, 1. , . 25, 0. , 3)
CALL RECT(XPAGE+1. 5, YPAGE, . 5, . 75, 0. , 3)
CALL RECT(XPAGE+1. 25, YPAGE-. 25, 1. , . 25, 0. , 3)
CALL RECT(XPAGE+. 312, YPAGE+. 75, . 25, . 125, 0. , 3)
CALL RECT(XPAGE+1. 312, YPAGE+. 75, . 25, . 125, . 125, 0. , 3)
CALL PLOT(XPAGE+. 5, YPAGE-. 25, 2)
CALL PLOT(XPAGE+. 687, YPAGE-. 25, 2)
CALL PLOT(XPAGE+. 687, YPAGE-. 75, 2)
CALL PLOT(XPAGE+1. 062, YPAGE-. 75, 3)
CALL PLOT(XPAGE+1. 062, YPAGE-. 25, 2)
CALL PLOT(XPAGE+1. 25, YPAGE-. 25, 2)
CALL PLOT(XPAGE+. 5, YPAGE+. 75, 3)
CALL PLOT(XPAGE+. 687, YPAGE+. 75, 2)
CALL PLOT(XPAGE+. 687, YPAGE+1. 25, 2)
CALL PLOT(XPAGE+1. 062, YPAGE+1. 25, 3)
CALL PLOT(XPAGE+1. 062, YPAGE+. 75, 2)
CALL PLOT(XPAGE+1. 25, YPAGE+. 75, 2)
RETURN
END
```

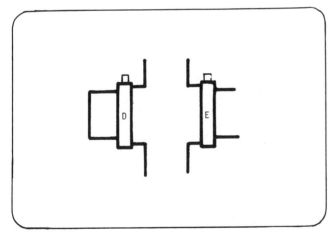

Figure 1.4 CAMD display of collar.

display of a keyway in the shaft and the pulley hub. This fixes the two
elements together by an element called a key. As shown in Figure 1.5,
the CAMD software routine RECT provides for these arrangements of
elements. Using this CAMD approach, pulleys may be free to slide along
their shafts while turning with them.

```
SUBROUTINE KEY(XSET,YSET)
CALL CIRCL(XSET+1.,YSET,206.,514.,1.,1.,0.)
CALL CIRCL(XSET+.5,YSET,(0.,360.,1.,1.,0.)
CALL RECT(XSET-1.5,YSET-.5,1.,1.,0.,3)
RETURN
END
```

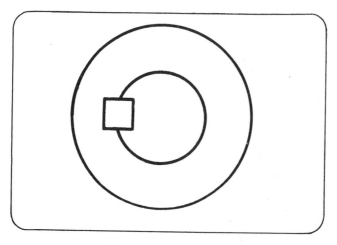

Figure 1.5 CAMD display of key.

The designer uses the display of pairs discussed earlier to create a sliding pair. This is often displayed by fitting to the elements a sliding key parallel to the axis of the shaft. The key may be made fast or may be an integral part of either element (shaft or pulley), or one of the elements may have a groove in which it can slide freely. This kind of display is very common, and is called a <u>feather and groove</u>.

Cranks and Levers

A <u>crank</u> can be displayed in a general way as an arm rotating or oscillating about an axis, as shown in Figure 1.6. When two cranks on the same axis are rigidly connected to each other, the name <u>lever</u> is often used to label software describing the combination of elements, particularly when the motion is over a small angle. The arms of a lever may have an angle between them; however, when the angle is less than 90° it is often stored in computer memory as a <u>bell crank,</u> and when the angle is more than 90° it is stored as a <u>rocker</u>: In machine design terms, this separation is often not used and the cranks are used interchangeably. When a computer is used to store basic elements of a machine, a separation must be made because the two lever arms may be in the same plane, or they may be attached to

```
      SUBROUTINE  CRANK
      NDATA = 250
      XTRANS = 2.5
      DO 5 I=1,NDATA
    5 READ(1,6) X(I),Y(I),Z(I),IPEN(I)
    6 FORMAT(3F6.3,I2)
C     DISPLAYS  FRONT  VIEW  OF  CRANK
      DO 10 J=1,NDATA
   10 CALL  PLOT(X(J),Y(J),IPEN(J))
C     DISPLAYS  SIDE  VIEW  OF  CRANK
      DO 20 K=1,NDATA
      Z(K)=Z(K)+XTRANS
   20 CALL  PLOT(Z(K),Y(K),IPEN(K))
      RETURN
      END
```

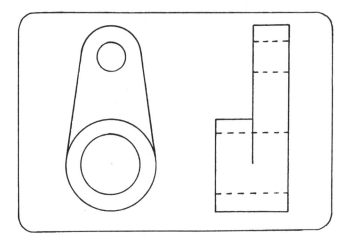

Figure 1.6 CAMD display of crank.

the same shaft but lie in different planes. In order to display all the combinations with their respective actions, this separation (bell and rocker)
is made.

Actions. A crank is displayed from computer database as a rigid element
connecting one member of a pair of cylindrical elements to one member of
another pair. The axis of one pair is assumed to be stationary, and the
axis of the other is constrained by the crank to move in a circular path
about the stationary axis.

Linkages.   A link may be displayed from a computer database as a rigid
element that serves to transmit force from one element to another. A linkage display consists of a number of pairs of display elements connected by
links. If the display combination is such that relative motion of the links is
possible, and the motion of each member relative to the others is definite,
the linkage becomes a kinematic chain. This type of display must be studied
at a graphics terminal that is capable of part animation (element motion).
If one of the links is fixed, however, the linkage becomes a mechanism.
Mechanisms may be studied by multicolor plotter outputs or DVST graphics
terminals.

       To study part animation requires the number of points whose motions
are determined by means of force links. Three links provide for a fixed
relationship, whereas four or more provide for movement, usually two
cranks. Although all of Chapter 5 is devoted to computer-aided linkage
design, it is important here to see how this display is made possible.

Slides and Slots.   Often when linkages are used in the computer analysis
of machine elements, a slot mechanism known as a slider-crank mechanism
is also employed. As indicated earlier, to analyze the motion of a machine
it is not always necessary to display in detail all its parts. Figure 1.7 is
the basic display of the essential elements of a slider-crank mechanism.

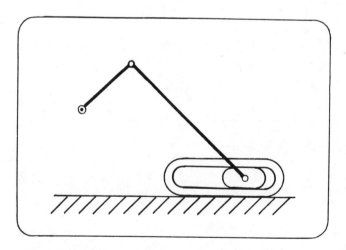

Figure 1.7 Slider-crank mechanism.

SUMMARY

The type of computer-aided design of machine elements described in this introductory chapter occupies the middle ground between tedious manual processes and supersophisticated computer methods. Such an intermediate process addresses the task of computer-aided machine design (CAMD) at a price that most small companies and universities can afford. The significance of the process is that a user does not have to be a computer science expert to apply the system. The machine designer will need to know FORTRAN and have access to a dual-processor interactive design system (DPIDS).

A DPIDS's concentrated processing power, diagrammed in Figure 1.1, is intended for the machine designer who needs a complete CAMD system at a reasonable cost. With this a designer is assured that the operation can be expanded into a small interactive network (Figure 1.3), a network that can link the engineer who designs the machine, to the element programmer who describes the member for display and analysis, to the CAM programmer who describes the part for manufacturing, to the technician who will inspect the finished product.

To accomplish this, a DPIDS contains two processors plus a host computer. One processor is dedicated to interactive design activity, and the second provides extensive design support and host file manipulations. Together, they form the basis of a powerful and economical interactive machine design system that delivers the following CAMD capabilities:

Geometric Construction
Design Documentation and Layout
    Two-dimensional or projective construction mode (Figure 1.2)
    Three-dimensional translation (Figure 1.4)
    Three-dimensional pattern placement (Figure 1.5)
    Level-sensitive pattern retrieval (Figure 1.6)
    User functions (Figure 1.7)

Detail generation
    Sectional views
    Dimensioning and rotation
    Threads and fasteners
    Piping details
Pictorial representation
Finite-element modeling
Animation technique for numerically controlled (N/C part
    programming

Figure 1.8 represents a typical CAMD workstation and Figure 1.9 illustrates the complete system of workstation, digitizer, plotter, processors, and page printer.

Figure 1.8 Typical CAMD workstation. (Courtesy of Auto-Trol Corp.)

Figure 1.9 Complete DPIDS for CAMD. (Courtesy of Auto-Trol Corp.)

APPENDIX—A BRIEF HISTORY
OF COMPUTER-AIDED MACHINE DESIGN

The evolution of computer-aided machine design can be traced along two
lines: that of the computer, a machine itself; and the history of machine
design. The history of machines is part of human history, for no other
being uses machines. A rather extensive treatment of machines and their
impact on society is given in A. P. Usher's work, A History of Mechanical
Inventions, published by Harvard University Press in 1929.

Usher divides the history of machines, including computers, into
pre-1857 and post-1857 sections. Credit for creating this major division
is given to Franz Reuleaux, in his 1857 publication, Principles of Machine
Design. Usher described the division as follows:

1. Pre-1857 machine designers generally accepted the existence of
   a small number of simple machines, which act in combination to
   form compound machines.
2. Post-1857 machine designers thought of a machine as a combination of elements described by Reuleaux as resistant bodies
   arranged such that the mechanical forces of nature are compelled
   to do work.

Reuleaux is probably best known for his 1876 publication Kinematics
of Machines and his theories of machine motion. This was his second
major work and is secondary to this discussion. What is important to note
is the work prior to 1857 of George Boole (1854, Laws of Thought), which
describes logic algebra, and that of Charles Babbage. Babbage's
"difference machine" (1812) did not directly influence the design of
computers or machine parts, but certain basic ideas of the stored
program can be traced to this remarkable nineteenth-century inventor.

Another pre-1857 designer, Joseph Jacquard (1745), devised a
method for using holes in cards to control the selection of threads in
weaving designs. Babbage used Jacquard's idea of holes in cards to
input data into his analytical engine. The post-1857 inventor Herman
Hollerith used both ideas to formulate his plan for the modern machine-
readable punched card. Hollerith realized the need for such a mechanical
tabulating device while conducting the 1890 Census. In 1896, he organized
the Tabulating Machine Company to manufacture and market the machines
and cards. This company merged with several others to eventually become International Business Machine Corporation (IBM).*

In this same general period, important works on machine design
were being published:

---

*The IBM 3033 has been used as the host computer for all the CAMD
examples shown in this book.

1870 R. W. Willis, Principles of Mechanics
1878 R. A. Proctor, A Treatise on the Cycloid
1883 F. Grashof, Theoretische Maschinenlehre
1886 Kennedy, A. Mechanics of Machinery
1891 Routh, E. J. Dynamics of a System of Rigid Bodies

James Powers, also employed by the Census Bureau, modified
Hollerith's card-processing equipment to tabulate the 1910 Census. His
improvements were so successful that in 1911 he formed the Powers
Accounting Machine Company, which was later purchased by Remington
Rand. Today Remington is a division of Sperry-Rand, the UNIVAC
manufacturer. During this development of the computer, machine
designers were investigating various fields:

1900 H. Hertz, The Principles of Mechanics
1912 A. G. Webster, Dynamics of Particles and of Rigid, Elastic,
          and Fluid Bodies
1917 A. W. Klein, Kinematics of Machinery

In 1937, Howard Aiken of Harvard University designed and IBM
began to construct a machine that would automatically perform a
sequence of arithmetic operations. Work was completed in 1944 on
the huge mechanical calculator, called Mark I. It contained 72
accumulators and 60 sets of switches for constants. Instructions
were given by means of switches, buttons, wire plugboards, and
punched tape. The Harvard Mark I was completed one year before
professors Eckert and Mauchly of the University of Pennsylvania,
working with Sperry, could finish the ENIAC.
    Using some electrical components, the ENIAC was much faster
than the Mark I. Neither machine stored any memory; each was programmed by external methods. Both IBM and Sperry began work on
the use of binary numbers for electronic arithmetic operations and the
internal storage of instructions written in digital form. The EDVAC
(Eckert and Mauchly) and the IAS (John von Neumann) introduced the
basic designs for two important types of computers (serial and
parallel). In 1946, the Eckert and Mauchly Computer Company produced the UNIVAC I, and in 1949 Sperry purchased the market rights.
Von Neumann, a noted mathematician working at the Institute for
Advanced Study (IAS) in Princeton, built upon the work done by
Eckert and Mauchly and completed the IAS computer in 1952.
    Had Usher written A History of Mechanical Inventions in 1952,
he might well have placed his dividing line at 1948. For in 1948,
Bell Laboratories developed the transistor. The transistor, an
electronic device, replaced the vacuum tube, an electrical device;
although neither is purely mechanical, the transistor is smaller

Table 1.4   Milestones in the Development of Computer-Aided Machine Design

Date	Event
1812	Babbage invents the difference engine
1854	Boole's Laws of Thought
1857	Reuleaux develops principles of machine design
1870	Willis develops principles of mechanics
1878	Protector writes a series of papers developing mathematical models for machine design
1883	Grashof develops the study of linkages
1886	Kennedy describes the theorem of three centers
1890	Hollerith uses punched cards for census
1891	Routh describes the dynamics of rigid bodies
1900	Hertz refines the principles of mechanics
1912	Webster expands existing theory to elastic and fluid bodies
1917	Klein develops the graphical kinematics method
1928	Lamb refines the statics of machines
1932	Root studies the dynamics of shafts and engines
1937	Aiken proposes the MARK I
1943	Eckert and Mauchly work on ENIVAC
1945	ADVAC and IAS projects begin
1948	Invention of the transistor
1951	UNIVAC I
1953	IBM delivers the first business computer
1954	First installation of UNIVAC for nonmilitary use
1956-1959	First second-generation computers installed
1963	First third-generation computers delivered
1965	Development of computers using integrated circuits
1969	Dual-processor minicomputers are commonplace
1975	Microprocessors are commonplace
1980	Interactive processors are commonplace in machine design processes

and less expensive, generates less heat, and requires less power. This really important development was applied to computers, reducing their size and operating environment. Now computers began to appear in manufacturing plant assembly areas, drafting rooms, and machine designers' work areas. The change in computers was so decisive that tube machines were referred to as first-generation computers. In 1953, IBM installed its first computer, the 701. In 1954, an improved model, the 650, was introduced. During the next five years, IBM introduced its second generation of computers. The changeover from first to second generation began in 1956 for military and in 1959 for commercial users. The second generation, using transistors, lasted until miniaturization techniques resulted in smaller, faster, and more reliable components for computers. Transistors, diodes, resistors, and other devices were packaged into modules. The computer circuitry was then made up of an arrangement of modules mounted on cards. This lead to the direct use of small, powerful computers in manufacturing (N/C) areas and in quick documentation of drafting and design needs. This third generation of computers roughly spanned the years 1959 to 1969, and introduced the dual-processor minicomputer discussed throughout this book. At this time computers were used predominately in the electromechanical and electronics design areas. Slowly, computer-aided drafting, design, and manufacturing applications began to emerge. But the evolution was not yet complete. Microelectronic techniques ushered in another generation: the integrated circuit was packaged to produce a microprocessor chip. This chip contained all the functions of the IAS or EDVAC computer in a component package small enough to hold on a thumbnail.

Table 1.4 lists some milestones in the development of computer-aided machine design.

The new generation of microcomputers has made "smart" terminals appear at the machine designer's desk. Coupled with a large host computer, the machine designer of today may select from:

IMP (Integrated Mechanisms Programs)
DRAM (Dynamic Response of Articulated Machinery)
MEDUSA (Machine Dynamics Universal System Analyzer)
DYMAC (Dynamics of Machinery)
VECNET (Vector Network)
ADAMS (named for inventor at University of Michigan)
AAPD (Automated Assembly Program)
UCIN (University of Cincinnati Kinematics Program)
CUCAMD (Clemson University Computer-Aided Machine Design
    software package)

BIBLIOGRAPHY

Anderson, R. H., Programmable automation: the future of computers in
    manufacturing. University of Southern California, Report No.
    ISI/RR-73-2, 1973.
Brown, N. L., Using a computer-aided graphics system to help design and
    draft auto components, Proceedings of the 14th Design Automation
    Conference, New Orleans, La., 1977.
Church, J. L., The ID of Current CAD/CAM Research in Product Design
    and Manufacturing, Illinois Institute of Technology and National Science
    Foundation, Washington, D.C., HT, NSF, April 1976.
Evans, D. C., Graphical man/machine communications, University of Utah,
    Report No. 5, 1968.
Evans, J. M., Strategies for modular CAD/CAM systems, Proceedings of
    the 15th Numerical Control Society, Chicago, 1978, pp.29-38.
Franz, D., A method for a CAD of variants, IBM Nachr., Vol. 28,
    No. 239, 1978.
Hack, H. S., CAD of manufacturing and erection plans, Symposium on
    Computers in Design of Chemical Plants, 1975.
LaCoste, J. P., Communications in CAD/CAM, Proceedings of the 2nd
    IFIP/IFIC, Budapest, 1973.
Laurer, D. J., CAD/CAM interactive graphics systems designed by users,
    Proceedings of the 15th Numerical Control Society, Chicago, 1978.
Ryan, D. L. Computer-Aided Graphics and Design. Marcel Dekker,
    New York, 1979.

2

Computer Display of Machine Motion

The purpose of computer displays for dynamic illustration of machine elements is to improve the computer-aided design of manufactured products by American industrial firms. This improvement normally involves those processes that require a description of geometric edges. The edges are joined together to form surface descriptions, while multisurfaces describe three-dimensional objects.

Machine designers have not used straight lines or circular arcs to define geometric edges. These two simple geometric entities (circle and line) allow little opportunity for aesthetic expression of functional engineering form design. From typewriters to rowboats, the typical manufactured product contour is free-flowing or "sculptured." Therefore, they are difficult to define and draft manually. The use of the computer has been of significant assistance. The automotive companies and the major aerospace firms have developed computer-aided design programs that deal with part animation, simulation, and documentation. But these programs are costly and jealously guarded. Several of these programs represent over $1 million each in development cost.

It is my contention that there is no reason why all computerized approaches must be expensive. Not all design problems are on the scale of an automobile or airplane. If the scale is modest, as in the case of a single machine element, an equally modest computer display of a machine motion program can be developed to handle it.

ANIMATION OF MACHINE PARTS

Just as significant portions of applied research have been developed "in-house" by giant corporations and major aerospace companies, medium-sized and smaller manufacturing companies are expending considerable efforts in the computer-aided machine designs (CAMD)

area. Those that have considerable research and development (R&D)
commitments (several millions) are all large-computer-center operations. Only a small number of manufacturers have explored the
possibility of small computer systems as described in Chapter 1;
these companies are:

Adage	Gerber Scientific
Applicon	Imlac
Auto-Trol	Evans and Sutherland

To date, only two provide a CAMD-DPIDS arrangement. This arrangement provides:

1. Speed; since DPIDS processing units handle the systems
   calculations, communications, and management routines,
   the workstation processor responds immediately to user
   design requests.
2. Reliability of multiprocessor architecture builds a high degree
   of hardware redundancy into a DPIDS.
3. Simplicity of DPIDS operating software is sensibly designed for
   high-speed execution and easy user understanding. Since each
   design station processor is devoted to just one user—not to the
   communication and application needs of a number of users—CAMD
   software is specific and concise.
4. This economy of computer memory is rapidly cost decreasing.
   Since CAMD utilizes a number of computer processors—not a
   single large host computer as in nondistributed processing
   systems—CAMD expansion is extremely economical.
5. The CAMD system is programmed in FORTRAN, the programming
   language for engineering applications. The hardware is designed
   with a human-engineered tutorial approach for ease of hardware/
   software interfacing and guides the user from start to finish of a
   product design.
6. Network expandability of a CAMD allows a company to expand,
   easily and economically, into an organized network—a network
   that gives every designer access to a common database and
   storage system resources. A network allows designers,
   detailers, part programmers, and inspection personnel to
   work side by side in groups of four in a total cluster of 64 users.

Path Programming

Part animation implies part motion. Motion is change of position. Motion
and rest are relative terms within the limits of CAMD. A part may be
conceived to be fixed in space, even though such does not exist in nature.
If two machine elements, both moving in space, remain in the same position relative to each other, they are displayed at rest, one relative to the

other. But if the two elements do not remain in the same position, either may be displayed in motion relative to the other.

The path program for motion may therefore be either relative or absolute (provided that some point is assumed to be fixed). Ordinarily, in nature, the earth is assumed to be at rest and motions referred to it are considered as absolute. On a computer display screen a point moving in space describes a line called its path, which may be rectilinear or curvilinear. The motion of a member is determined by the paths of three of its points not on a straight line. If the motion is in a plane, two points suffice; and if rectilinear, one point suffices to determine the motion. Figure 2.1 illustrates a path program.

Direction or Sense

If in a machine part animation display, a point is moving along a straight path, the direction of its motion is along the line that constitutes its path. Motion forward toward the end of the path is assumed to have a positive direction and is indicated by a + sign, whereas motion backward toward the start of the path would be negative and indicated by a - sign. In dynamic displays this is often referred to as the sense of the display motion. For example, if a point moves along a straight line in the path programming output shown in Figure 2.2 from starting point A toward ending point B, the sense of the motion is from A toward B, or simply AB. If a point is moving along a curved path, the direction at any instant is along the tangent to the curve, as illustrated by the subprogram in Figure 2.3.

Continuous Motion

When a point continues to move indefinitely in a given path in the same sense, its motion is displayed as continuous. In this case the path must return on itself, as in a circle or closed curve. In Figure 2.3 continuous motion would be the result if SANG = 0. and N*THETA = 360. In this case a circle would be described starting at the X, Y, and Z locations in space. The size of the path is determined by the value of R (radius) of the path.

Reciprocating Motion. When a point traverses the same path and reverses its motion at the end of that path, the point motion is displayed reciprocating. In reciprocating machinery, motion consists of three display elements: displacement, velocity, and acceleration. Display motion is described in this chapter. Chapters 3 and 4 deal with velocities and accelerations, respectively. The displacement of a point on an animated display of a machine part may be according to a definite path such as the circle path in Figure 2.3, and is therefore expressed in equation form. If the equation for motion exists, the translation of that equation to a set of computer instructions (FORTRAN) is a rather

```
C  *  ****************************************************************
C  *                                                                *
C  *  PART ANIMATION PROGRAM FOR DISPLAYING MACHINE ELE-            *
C  *  MENTS AS WIREFORM MODELS.  EACH MODEL IS CONSTRUCT-          *
C  *  ED BY DEFINING PART POINTS.  THE POINTS ARE MOVED IN         *
C  *  SPACE AND THEN CONNECTED FOR VISUALIZATION PURPOSES.         *
C  *  THE FOLLOWING INPUT INFORMATION IS REQUIRED FOR              *
C  *  EACH PART:                                                   *
C  *                                                                *
C  *           NP = NUMBER OF PART POINTS TO BE DISPLAYED          *
C  *           NC = NUMBER OF POINT MOTIONS REQUIRED TO            *
C  *                CONNECT                                        *
C  *           NV = NUMBER OF WIREFORMS TO BE DISPLAYED           *
C  *        SPACE = TRANSLATION FOR EACH WIREFORM                 *
C  *                (IF SPACE = 0.  ABSOLUTE DISPLAY RESULTS)     *
C  *        P(I,J) = ARRAY CONTAINING 3-D DATABASE FOR            *
C  *                EACH POINT                                     *
C  *      VP(NV,3) = ARRAY CONTAINING AMOUNT OF ROTATION          *
C  *                                                                *
C  *  ****************************************************************
       DIMENSION P(100,3), IC(200), VP(100,3), PP(100,3)
       READ(1,1)NP,NC,NV,SPACE
     1 FORMAT(3I5,F6.3)
       READ(1,2)((P(I,J),J=1,3),I=1,NP
     2 FORMAT(3F6.3)
       READ(1,2)((VP(I,J),J=1,3),I=1,NV
       READ(1,3)(IC(I),I=1,NC)
     3 FORMAT(10I5)
       CALL PLOTS
       DO 5 I=1,NV
       A=ARTAN(VP(I)),VP(I,J)
       SA=SIN(A)
       CA=COS(A)
       DO 6 J=1,NP
     6 PP(J,1)=P(J,1)*CA-P(J,3)*SA
       VPP=VP(I,3)*CA-PC(J,3)*SA
       A=ARTAN(VP(I,2),VPP)
       DO 7 J=1,NP
     7 PP(J,2)=P(J,2)*CA-PP(J,3)*SA
       DO 11 K=1,NP
       PP(K,2)=PP(K,2)+6.
    11 PP(K,1)=PP(K,1)+SPACE*I
       DO 8 J=1,NC
       IF(IC(J).LT.0.)GOTO9
       CALL PLOT(PP(IC(J),1), PP(IC(J),2),2)
       GOTO8
     9 K=IC(J)
     8 CALL PLOT(PP(K,1),PP(K,2),3)
     5 CONTINUE
       CALL PLOT(SPACE*NV,0.,999)
       STOP
       END
```

Figure 2.1   Path programming.

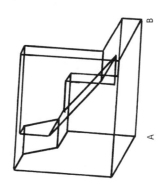

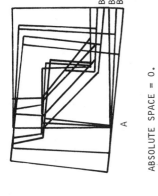

ABSOLUTE SPACE = 0.

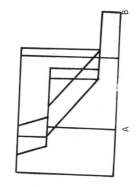

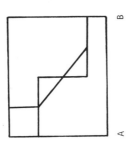

Figure 2.2   Path program output.

```
C  *  ********************************************************************
C  *                                                                    *
C  *  PART ANIMATION SUBROUTINE FOR DISPLAYING A CURVI-                 *
C  *  LINEAR PART MOTION.  THIS SUBPROGRAM PRODUCES                     *
C  *  DATABASE FOR USE IN THE PART ANIMATION PROGRAM.                   *
C  *  THE FOLLOWING INPUT INFORMATION IS REQUIRED:                      *
C  *                                                                    *
C  *          X Y Z = STARTING POINT LOCATION                           *
C  *            R = RADIUS OF POINT PATH                                *
C  *         SANG = STARTING ANGLE ALONG POINT PATH                     *
C  *            N = NUMBER OF ARC SEGMENTS IN PATH                      *
C  *        THETA = SIZE IN DEGREES OF ARC SEGMENTS                     *
C  *                                                                    *
C  *  ********************************************************************

         SUBROUTINE CURVE(X,Y,Z,R,SANG,N,THETA)
         SANG=(3.14/180.)*SANG
         P=X-R
         XX=R*(1-COS(SANG))
         YY=R*(SIN(SANG))
         DX=P+XX
         DY=X+YY
         WRITE(3,3)DX,DY,Z
         THETA=(3.14/180.)*THETA
         THETA1=THETA
         D0 2 I=1,N
         FEE=SANG+THETA
         PX=R*(1-COS(FEE))
         SY=R*(SIN(FEE))
         DX=P+PX
         DY=Y+SY
         WRITE(3,3)DX,DY,Z
     2   THETA=THETA + THETA1
     3   FORMAT(3F6.3)
         RETURN
         END
```

Figure 2.3   Curvilinear part point motion.

straightforward process. Often, the displacement is not easily defined.
In these cases an interactive graphical display is useful. A line display
of the essential elements of the machine is drawn on the graphics tablet
and input to the CAMD-DPIDS in various positions. The desired points
are located at each position, and a program is written to display a graph
to show the displacement. In most cases, only the extreme positions are
needed. This procedure not only provides an accurate displacement of a

part, but more important, it offers a means of visually determining the
interference of parts in a machine.

Oscillation.   Oscillation is a term applied to reciprocating circular
motion, such as that of a pendulum. As indicated for reciprocating
display motion, it is not necessary to display in detail all the parts of
a machine in order to study oscillation. Figure 2.4 is an example.
Display points FA(2) and FA(4) are fixed axes. FA(2),A = 1.5 display
units, A,B = 3., FA(4),B = 2., and FA(2), FA(4) = 3. units. A display
unit may be set equal to an English inch or a convenient metric size.
American manufacturers of computer display hardware supply either
method of scaling. Scaling does not mean the conversion from English
to metric, however. Scaling is the conversion from 1024 possible display positions across a CRT screen, for example, to a more easily
understood unit size. The scale is selected to match the physical limits
of the display device.
         Crank 2 in Figure 2.4 is the driver turning counterclockwise. The
display is such that, while 2 makes a complete revolution, 4 oscillates
through an angle. By the computer animation technique the reciprocating
member 2 and oscillating member 4 can be visually checked for displacement.

Intermittent Motion

When the motion of a point is interrupted by periods of rest, its motion is
displayed as intermittent. Figure 2.5 is an example; here block 4 slides

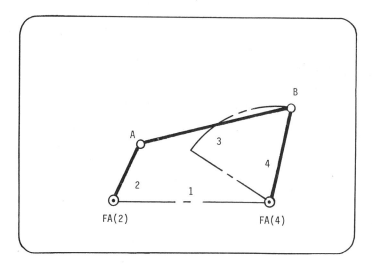

Figure 2.4   Reciprocating/oscillation display.

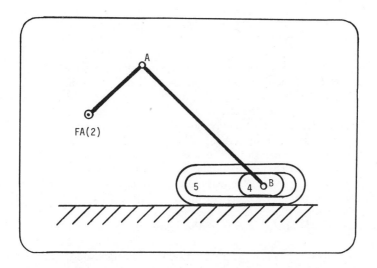

Figure 2.5   Intermittent motion display.

inside member 5, which slides along a surface related to the crank FA(2).
The crank acts as a driver turning counterclockwise. The display is such
that, while 2 makes a complete revolution, AB oscillates, causing block 4
to slide inside member 5 and member 5 to slide along the indicated surface.
Although the motion of block 4 is not interrupted, member 5 has a period of
rest in its motion.

Rotation and Revolution

A point is displayed to revolve about an axis when it describes a circle of
which the center is in the axis and the plane of rotation is perpendicular to
that axis. Figures 2.4 and 2.5 both contain a rotating crank element.
When all the points of an element move, the element is displayed as a
revolution about the axis. If this axis passes through the center line of
the element, as in a wheel, the display is both a rotation and a revolution.
The mechanical engineering term turn is used synomymously with either
rotation or revolution. It frequently happens in machine design that an
element not only rotates about an axis passing through it, but also moves
(revolves) in an orbit about another axis. In CAMD the two display techniques are visually different.

Axis and Planes.   An axis of rotation is a line whose direction is not
changed by the rotation; a fixed axis is one whose position and direction
remain unchanged. A plane of revolution is a plane perpendicular to the
axis of rotation.

<u>Directions and Cycles</u>.   The direction of rotation or revolution is defined and noted on a computer display by labeling the direction of the axis. The sense is labeled by stating whether the turning is right-handed (clockwise) when viewed from a noted side of the plane of motion.

When a mechanism is set in motion and its parts go through a series of movements that are displayed over and over while the relationship between and order of the animated segments of the display series remain the same, a cycle of motion is displayed. This animation of machine parts is called the <u>kinematic cycle</u>.

<u>Periods</u>. A <u>period</u> of motion is the time occupied by the animation series in completing one cycle. Periods, cycles, rotation, revolution, oscillation, reciprocation, sense, direction, and path programming are the necessary physical entities that must be displayed for successful animation of machine parts. All the concepts presented thus far in this chapter refer to the display portion of CAMD, the ability to present concepts as visual-graphical entities for study and analysis by the machine designer. The designer may interact with the visual display and change entities at will, but very little computer-aided design (CAD) capability has been presented. Animation deals almost entirely with the computer-aided graphics (CAG) portion of CAMD (see Figure 1.1). To solve design problems related to velocity and acceleration, a simulation program is developed.

SIMULATION OF MACHINE PART MOTION

Since the motion of an element is determined by the animation of not more than three of its component particles, not laying in a straight line, it is essential before beginning the simulation of the motion of rigid machine elements that the laws governing the simulation (animation plus analysis) motion of a point be fully understood.

Linear Speed

<u>Linear speed</u> is the time rate of motion of a point along its path during animation. In the path program it is the rate at which a point is approaching or receding from another point in its path. If the point to which the display of the moving point is used as a reference, the speed is the absolute speed of that point. If the display point is itself in animation, the speed of the point is relative. Linear speed is expressed in linear display units per unit of time.

Angular Speed

<u>Angular speed</u> is the time rate of turning the displayed element about an axis, or the rate at which a line on a revolving machine element is changing direction. Both are expressed in angular display units per

unit of time. If a display is revolving about an axis, any point in the display has linear speed only. But a line, real (boundary) or imaginary (center line), joining that point to the axis of revolution has angular display speed. A line joining any two points on the display to form the wireform has angular speed, also.

Uniform or Variable Speed

The display speed used in simulation is uniform when equal distances are presented in equal times, regardless of the time rate. The speed is variable when unequal distances are presented in equal intervals of time. When only the animation is considered, the acceleration of a point is zero. The speed is uniform (the moving point is displayed over equal distances in equal intervals of time). The speed (velocity) is therefore equal to the length of the path L, in linear units, divided by the time T, in time units required to travel the path. This is expressed in a simulation program as

REAL L
$V = L/T$

where V is stored in linear display units per unit of time. The scale of the linear display units of V and L must be the same, and the time rate for display of V must be the same as T.

Velocity

The term velocity is used synomymously with speed by machine designers. This is incorrect in a CAMD simulation program, because velocity includes direction and sense as well as speed. The linear velocity of a point is not fully identified for display until the direction and sense in which it is moving and the rate at which it is moving are found in the program. Therefore, velocity can be programmed as either uniform speed plus direction and sense or variable speed plus direction and sense.

Acceleration

Simply stated, acceleration in a simulation program is the rate of change of linear or angular velocity. For simulation purposes it is convenient to classify point motion according to the kind of acceleration the point displays:

1. Acceleration zero (constant velocity).
2. Acceleration constant.
3. Acceleration variable according to a simple computer program statement expressed terms of V, L, or T in a CAMD simulation program providing CAG, CAD, or CAM output.

Linear Acceleration.   Since velocity involves direction as well as rate of motion, linear acceleration may involve a change in speed or direction or both. Any change in the speed of the computer display takes place in a direction tangent to the path of the display point and is called <u>tangential acceleration</u>. A change in display direction takes place normal to the display path and is called <u>normal acceleration.</u> Given this, acceleration may be either positive or negative. For example, if the speed is increasing, the acceleration is positive. However, if the speed is decreasing, the acceleration is negative and is called <u>deceleration.</u>

<u>Angular Acceleration.</u>   As in linear acceleration, a change in either display speed or direction of rotation or both may be involved. <u>Angular acceleration</u> in a simulation program is understood to refer to a change in angular speed. Angular acceleration is expressed in angular units, change of speed per display unit of time (such as radians, degrees, or revolutions per time). Most simulation programs require the use of radians, which is the angle subtended by the arc of a circle equal in length to its radius. Since the radius is contained in the circumference 2π times, there must be 2π radians in 360° or 1 radian is equal to 57.296°.

   A conversion from degrees to radians was necessary in Figure 2.3. The machine designer inputs the starting angle along the point path in degrees. In order to display the output in dynamic form, the storage location called SANG was changed from degrees to radians. The basic formula,

   2π radians = 360

was written as

   $$RADS = \frac{360}{2\pi} \qquad \text{or} \qquad SANG = SANG*(3.14/180.)$$

Translation

A display from a simulation program has motion or <u>translation</u> when all its elemental points have the same velocity. That is at that instant, all points are moving in the same direction with equal speeds. In that case the acceleration would be constant.

   Let A, from Table 1.1, represent the acceleration and the number of display units (speed) added per unit of time T. Then during T the change in speed is A*T. If at the beginning of that display time interval, the speed is V(0), then at the end of T the speed will be V(0)+A*T. Therefore,

   $V=V(0)+A*T$

From this the average display speed is

   $$\frac{V(0)+(V(0)+A*T)}{2} \qquad \text{or} \qquad V(0)+.5*A*T$$

and since the distance traveled in the display is the average speed multiplied by the time, then

   $L = V(0)+(.5*A*T)$

DOCUMENTATION OF MACHINE MOTION AS COMPUTER OUTPUT

The display of an animated machine element operating under normal conditions is called <u>simulation.</u> Simulation involves the acceleration of a moving particle and may vary as some function of distance moved, velocity, or time. When certain machine conditions exist, definite equations may be written expressing the relationship among A, L, V, and T.

Three cases will be considered:

1. Harmonic
2. Parabolic
3. Uniform

Often no direct relation exists among acceleration, velocity, distance moved, and time that can be conveniently expressed in the form of equations. The database for the part simulation may be obtained by observations or computations at certain frequent intervals in the simulation program. These intervals correspond to the cycle of motion and the animation worked out on graphical displays.

The process of displaying simulation problems of this type consists of approximating, by means of animation graphics, the necessary differentiations or integrations instead of solving them directly from the equations. Small display increments ΔL, ΔV, and ΔT are used instead of the infinitely small dl, dv, and dt. Then, where differentiation is required, the ratio of dl/dt or dv/dt is found from counting display points on the output memory device. Points on the display area range from 0 to 1023 for X and 0 to 761 for Y. Point $(0,0)$ is located at the bottom left of the display screen, as shown in Figure 2.6.

Where integration is involved, the equivalent is obtained by the summation of the finite increments found on the display area. This is expressed by Σ instead of the integral sign, $\int$. For example, $\Sigma\Delta V$ means the total sum of the successive values of ΔV. Although this procedure is extremely easy to document in a simulation program (FORTRAN), the reduced amounts of data provide a jerky display motion.

Similarly, in the use of the equations

$$V = \int_{T(0)}^{T} A \, dt \quad \text{or} \quad V = dl/dt$$

and

$$dt = dl/V \quad \text{therefore} \quad T = \int \frac{dl}{V}$$

a huge database is calculated and stored by the simulation program and all the points are displayed at the same time, causing the display area to become overloaded. Overloading on a refresh CRT causes display flicker because the display processor cannot cycle through the long list in the

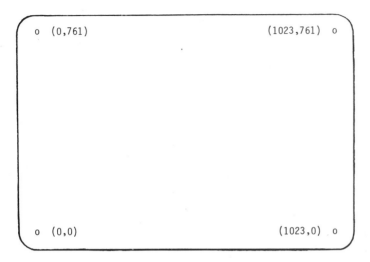

Figure 2.6   Display points.

time required for display (1/30 sec). More data are computed than
necessary, causing the long list; to refresh the animated display 30
times a second, a data-reduction subroutine must be employed. This
increases the host processing time, making the economics questionable.
Overloading of a DVST terminal causes "hot spots" on the inside surface
of the display tube and constant burning away of tiny particles. Dead
areas in CRT display screens are common from improper software
approaches.

Summation routines cause jerky displays; they are very inexpensive,
but cause eye strain and fatigue. Simulation programs without data reduction are medium range in cost but overload the display area, again causing
eye fatigue. The ideal solution is to group the simulation motion desired.

Harmonic Motion

Into this group are placed types of motion in which the acceleration varies
directly as the displacement is known. These are known as simple harmonic
motions. The most common example is reciprocation over a straight path,
with the sense of the acceleration always toward the center of the path. The
nature of the motion may be visualized by reference to Figure 2.3. Suppose
that a display point moves with uniform speed around the circumference of a
semicircle of radius R, center location at X, Y, Z, and diameter P to X+R.
Assume further that a display beam moves along the diameter at such a
speed that it is at all times at the base of a perpendicular dropped from
THETA to N. If this beam moves with a linear speed V(P), the radial line
will turn at an angular speed equal to V(P)/R. This constant angular speed

from Table 1.1 has been called OMEGA. The OMEGA = Π divided by the
time required for the display from P to X+R in Figure 2.3. If V(P) = velocity
of P, then using the FORTRAN statement

$$V(P) = OMEGA*R*(SIN(THETA))$$

the harmonic motion can be simulated.

Parabolic Motion

Three cases of parabolic motion are allowed:

1. A = a function of T
2. A = a function of V
3. A = a function of L

Since the display acceleration is the time rate of change of velocity, if this
rate of change in a time ΔT is constant, then A = $\Delta V / \Delta T$, and as ΔT is
decreased, this approaches a limit dv/dt. Hence, A = dv/dt at any display
instant.

By the use of these equations of motion, problems involving the afore-
mentioned three cases can be displayed. In some cases, especially where
V has some value other than zero, the resulting equations may be awkward
to convert to FORTRAN. It would then be advisable to resort to the Σ
methods described earlier.

Uniform Motion

The third type of motion is uniform motion. Instead of causing an
animated machine piece to travel its entire path with variable motion,
it is sometimes desirable to have it travel its path with uniform motion.
For example, a body is to start from rest at A and move to A(1) in time
T, accelerating uniformly over the distance M at the beginning, moving
with constant speed over the distance D, and decelerating uniformly
over distance M at the end (A(1)). T will represent the time required to
move the distance M with constant acceleration A, and V the speed at the
end of time T. Then either

$$M = .5*A*T**2 \quad or \quad A = 2*M/T**2$$

may be used in a simulation program and displayed as shown in Figure 2.7.

SUMMARY

This chapter described the animation, simulation, and documentation
techniques in CAMD. These advanced techniques can be broken down
into two major areas, application and systems. The application program

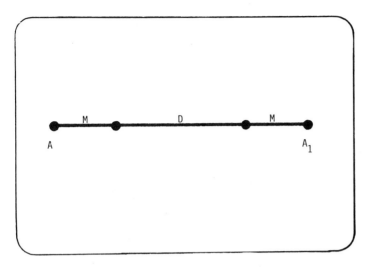

Figure 2.7   Uniform motion.

makes requests to the DPIDS for graphic output. Before actual display,
the graphic output is manipulated by graphic transformation. Basic trans-
formations include rotation, scaling, clipping, and windowing. The display controller diagrammed in Chapter 1 supports a transformed display
file, maintaining separate display code sequences for each individually
identified refreshed picture entity. Any graphic output request that is
not identified with a refreshed object is directed between cycles to the
face of the storage display. This allows distinct graphic entities to be
manipulated independent of one another.

Animation of machine parts is now possible based on the system just
summarized and the execution of a path program. The path program
describes the animation sequence: number of points, motion cycles,
display forms, and data arrays. The direction or sense of a display
point along its path is controlled by subroutines such as the one shown
in Figure 2.3. Various types of display point motion can be programmed,
such as continuous, reciprocating, oscillating, intermittent, and axis
revolution.

Animation plus the analysis of point motion yields machine part
simulation. Simulation is a technique the machine designer employs to
model the proposed part and test part speed/velocity, acceleration,
deceleration, and translation. When the testing is complete, a part
file is created where the database is documented by file listings (hard
copy) and memory modules (core storage).

APPENDIX — TSO COMMANDS

The following is a list of TSO commands and a brief description of the use
of each. For further information on how to use these commands, type
HELP (or H) followed by one or more blanks, followed by the name of the
command or its alias. Aliases for commands appear in parentheses. An
* after a name indicates that the command is also subcommand of CEDIT.
To get help with these commands while under CEDIT, type XHELP
followed by one or more blanks followed by the command name.

$ COMMANDS * - a series of commands used to display information
        about the HASP (job) queue. Type HELP $ for more information.
ALIAS (AL) - used to put alias names on members of partitioned data
        sets that are allocated SHR (i.e., to more than one user or job
        at a time).
ALLOCATE (ALLOC) * - used to allocate or create temporary data sets,
        associate a file name with a data set, or designate files for output.
ATTRIB (ATTR) * - used to define DCB parameters to be used with a
        subsequent ALLOCATE command.
CALL - used to execute load modules (programs already compiled and
        link edited) at the terminal.
CEDIT (CE) * - invokes the editor.
CHANGE - used to change the name of a member of a partitioned data set
        that is allocated SHR (i.e., to more than one user or job at the
        same time).
CHARGES - used to display a list of TSO session statistics.
CHAT * - used to transmit messages to users at other terminals.
CODE - used to produce an explanation of system completion codes and
        FORTRAN error codes.
CONCAT * - used to concatenate (hook together) two or more data sets
under a single file name.
COPY - used to copy (to another data set) a sequential or partitioned data
        set or member of a partitioned data set, to add a member to an
        existing data set, or to merge two partitioned data sets.
CPS - used to invoke the Conversational Programming System, which
        allows interactive programming in BASIC and PL/I.
CREPRO - has the same function as REPRO except that the job card can
        be supplied in a clist (command procedure), whereas in REPRO
        it cannot.
CSCRIPT (CS) - invokes the Waterloo SCRIPT text formatter using as
        input a member of the Source Management System.
DELETE (D) - used to delete a sequential data set, partitioned data set,
        or a member of a partitioned data set.
EDIT (E) * - invokes the IBM editor.
EXEC (EX) - used to execute a clist (command procedure).

FETCH (F) * - HTIP command used to list a program or part of a program currently on the fetch queue at the terminal or to save it into a data set. (Note: F is not a valid alias for this command under CEDIT.)

FORMAT (FORM) * - used to provide formatting capabilities for text-oriented data sets.

FREE * - used to unallocate data sets and files allocated to the user's TSO session and to delete attribute lists.

FREEALL (FA) * - used to unallocate ALL data sets and files allocated to the user's TSO session except those allocated by the LOGON procedure. (Note: FA is not an alias for this command under CEDIT.)

HELPJCL - used to obtain a listing of the statements necessary to use a cataloged procedure.

HEX - used to do hexadecimal addition or subtraction at the terminal.

IDA - invokes the Interactive Data Analysis statistical system.

LABEL - used to insert up to four characters in the line number field of a data set.

LESSON - provides information about a series of computer-assisted lessons designed to teach the user how to use the editor and Source Management System.

LINK - used to execute the linkage editor from the terminal.

LIST (L) - used to list the contents of a sequential data set or a member of a patitioned data set.

LISTALC (LISTA) * - used to list data sets and files currently allocated to the user.

LISTATTR - used to list the attribute lists currently allocated.

LISTBC (LISTB) * - used to list messages saved in the broadcast data set.

LISTCAT (LISTC) * - used to list the names of cataloged data sets.

LISTDS (LISTD) * - used to list DCB attributes of a data set or the members of a partitioned data set.

LISTHEX * - used to list the contents of a data set in hexadecimal.

LISTMEM * - used to list members in the Source Management System.

LOADGO (LOAD) - used to execute the loader, which loads an object deck (compiled program), resolves external references, and executes the program, from the terminal.

LOGOFF - used to end a TSO session.

LOGON - used to initiate a TSO session.

MERGE (M) - used to combine data sets or parts of data sets or to copy a data set or part of a data set.

PDSLOOK - lists members of a partitioned data set to screen.

PRINT * - HTIP command used to print a job on the fetch queue.

PROCS - used to concatenate a data set containing clists (command procedures) to the clist data set allocated at LOGON.

PROFILE (PROF) – used to specify to the system certain user characteristics to be used to control the flow of information to and from the
terminal (such as whether or not the terminal is to receive messages
from other users) or to list the user profile.

PURGE * – HTIP command used to delete a job from one of the HASP
queues.

RENAME (REN) – used to change the name of a sequential data set, a
partitioned data set, or a member of a partitioned data set, or to
create an alias for a member of a partitioned data set.

REPRO * – HTIP command used to have the contents of a data set printed
or punched.

SCHEDULE (SCHED) * – HTIP command used to submit a source program
from the terminal.

SCRATCH (SCR) – used to delete a member of a partitioned data set that
is allocated SHR (i.e., to more than one user or job at the same
time).

SCREPRO – has the same function as CREPRO except that it prints or
punches a member of the Source Management System.

SEND (SE) * – used to send messages to users at other terminals.

SHOW (STATUS, ST) * – HTIP command used to show the status of a
particular job, status of ALL user jobs on HASP queues, number
of jobs awaiting execution, different output files for a job, or
number of sers logged on.

SIZE * – used to list the amount of space allocated to a data set and how
much of that space is used.

SLIST (SL) – lists a member of the Source Management System.

SPACE * – used to produce a list of the disk packs available to house
temporary data sets and how much space is available on them.

SPEAKEZ – used to invoke the SPEAKEASY processor.

SREPRO – same as REPRO except that it prints or punches a member of
the Source Management System.

SSUB (SS) – submits a job from a member of the Source Management
System.

SUBMIT (SUB) * – has the same function as SCHEDULE.

TERMINAL (TERM) – used to define terminal characteristics (such as
number of lines per screen).

TEST (T) – used to debug assembler language programs at the terminal.

TESTCOB (TESTC) – used to debug COBOL programs at the terminal.

TESTFORT (TESTF) – used to debug FORTRAN programs at the terminal.

TIMEDAT * – used to display the Julian date and the time of day.

TYPE * – used to have a message printed at the user's terminal.

UNNUM – used to remove the line numbering from a data set.

US * – used to display a list of users currently logged on or to display
the number of users logged on.

WHEN - used to test a system return code from a previously executed
command.
WHOGOT * - used to determine to whom or to what a particular data set
is allocated.

BIBLIOGRAPHY

Andrews, H. C., Computer, digital image processing, IEEE Spectrum,
May 1974.
Boguslavsky, B. W., Elementary Computer Programming in FORTRAN.
Reston Publishing, Reston, Va., 1974.
Chasen, S. H., Geometric Principles and Procedures for Computer
Graphic Applications. Prentice-Hall, Englewood Cliffs, N.J., 1978.
Cripps, M., An Introduction to Computer Hardware. Winthrop Publishers,
Cambridge, Mass., 1978.
Croft, F. K., Why Use computer graphics? Proceedings of the Computer
Graphics Seminar, Clemson University, Clemson, S.C., November
1978.
Giloi, W. K., Interactive Computer Graphics. Prentice-Hall, Englewood
Cliffs, N.J., 1978.
Hollingum, J., Computer graphics cuts down drawing board time,
Engineer, May 1977.
Paul, B., Kinematics and Dynamics of Planar Machinery. Prentice-Hall,
Englewood Cliffs, N.J., 1979.
Phelan, R. M., Fundamentals of Mechanical Design. McGraw-Hill,
New York, 1970.
Rembold, U., M. K. Seth, and J. S. Weinstein, Computers in manufac-
turing. Marcel Dekker, New York, 1977.
Ryan, D. L., Computer-Aided Graphics and Design. Marcel Dekker,
New York, 1979.
Spotts, M. F., Design of Machine Elements. Prentice-Hall, Englewood
Cliffs, N.J., 1978.

3

Computer-Aided Velocity Analysis
for CRT Displays

Chapters 1 and 2 have described how the computer-aided design of machine
elements is accomplished from an overall hardware and software system
approach. This chapter describes the technique in hardware and software
for the display of velocity analysis [intermixed display of refresh and
direct-view storage tube (DVST) storage graphics]. The concept has
been presented that traditional storage tube cathode ray tubes (CRTs) have
the advantage of low cost coupled with the ability to display large amounts
of graphic information. These displays do not provide for the dynamic
motion and transformation required in animation and simulation of machine
parts. High-resolution and density-refreshed CRTs tend to be very expensive and often require considerable software (advanced equation database). Combining both refresh and storage graphics on the same display
allows computer-aided machine design (CAMD) output to be separated into
static and dynamic areas to achieve the velocity and acceleration analysis
at reduced cost.

VELOCITIES IN COMPUTER-AIDED MACHINE DESIGN

Velocities in CAMD may be determined analytically (from software only)
or graphically (hardware and firmware). Often, an analytical analysis becomes quite complicated and, in some cases, impossible. The analysis by
hardware/firmware are more direct, less complicated, and usually sufficiently accurate. This is achieved by storing images directly on the face
of the DVST, which eliminates the need for software image regeneration.
Now, by adding a fast, constant-rate vector generator and a high-speed
deflection system, a hardware feature known as write-through is possible.
This feature displays flicker-free refreshed images together with stored
images.

Besides this special hardware feature, it is also necessary to provide integrated software and system support called <u>firmware.</u> The provision of firmware is the novel aspect of the CAMD system. The machine designer can construct "refresh" and "storage" pictures with the same commands and can use the CAMD operating system diagrammed in Figure 1.3. It consists of a minicomputer, random-access memory, a micro-coded display controller, and a 19-in. refresh/storage display unit. On command from the minicomputer, the display controller accesses the display list in memory via the high-speed direct-memory-access channel. The display list contains beam positioning and status information. The display controller directs the vector generation from which a graphical velocity diagram is constructed.

Display Vectors

There are three methods for displaying velocities: (1) resolution and composition, (2) instantaneous axis of velocity, and (3) centros. Each type of display has its advantages. Some CAMD problems may be solved easily by one method. As a general rule, display methods 1 and 2 are the quickest (and therefore the cheapest). Method 2 is a simplified version of 3 and can be used in the solution of practically all problems.

The smallest graphical entity in a velocity display is the display vector. A <u>vector</u> is a line image that represents a vector quantity. The length of the vector displayed at any convenient scale indicates the magnitude. The direction of the vector is indicated by an arrowhead showing the sense of the vector. The starting position of the vector is called the <u>tail</u> and the end is called the <u>tip.</u> The sense of the vector is from tail to tip, with the arrowhead placed at the tip.*

Line segments displayed on an output device were labeled "starting point" and "ending point" in Chapters 1 and 2 because they were not vectors. These displays were known as <u>scalar</u> line values; this chapter deals with vector line segments. Vectors use a form of notation called <u>Bows</u>. Scalar notation is illustrated in Figure 3.1, vector Bow's notation in Figure 3.2.

You will notice that the space around the vector has been labeled. Both display segments are labeled AB. Line segment AB has length only, whereas vector AB has length and sense. The length of vector AB is directly related to its magnitude, to some display scale, and the arrowhead shows the direction of magnitude.

Complete vector notation also includes an angle specification for direction. In this chapter the reference line for measuring this angle will be a horizontal line displayed through the point of application and the angle will be measured in the counterclockwise direction. In Figure 3.3, the vector AB has a sense of 45° and can be written as

*A complete description of display vectors can be reviewed by studying Ryan (1979, Chap. 8).

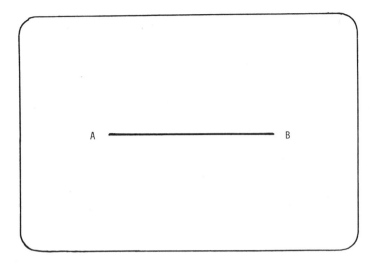

Figure 3.1   Scalar notation.

$$\overline{AB} = 150 \; \lfloor 45^o$$

In Figure 3.3, the display vector notation is used to express a vector, AB, which is 150 display units long (magnitude); the angle from the horizontal reference is 45°.

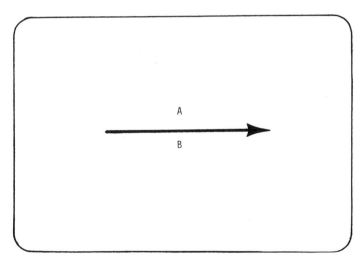

Figure 3.2   Bow's notation.

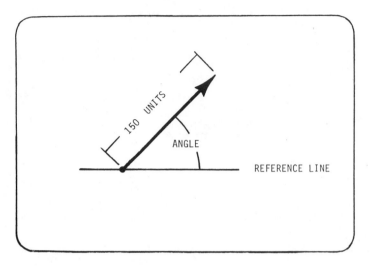

Figure 3.3   Vector notation.

If two or more display vectors are used in describing a velocity, a vector system is displayed on the CRT. To solve system problems, vectors are often combined to show the effect of the total quantities. Figure 3.4 illustrates the combination of two vectors.*

The combination of two vectors, displayed tip to tail, is called their sum or resultant. The sum of the two vectors in Figure 3.5 is a quantity whose effect is the same as the combined effect of the two original quantities. In Figure 3.5A the sum is the mathematical sum of the two quantities shown by the vectors themselves. In Figure 3.5B the sum of the two vectors is the closing side of a triangle whose two sides are the display vectors.

Similarly, a sum may be displayed representing the resultant of any number of display vectors. In Figure 3.6 the sum of the display vectors is the resultant, and its vector is called the resultant vector. The display vectors added together to obtain the resultant are its components. The sense of the resultant vector is toward the tip of the last display vector added. Figure 3.6 can be viewed on the display CRT by the following display program:

*Vector notation should not be confused with Bow's notation. Bow's is the labeling of a diagram for reading around two or more vectors called a system. The reading is made in a clockwise direction for Bow's.

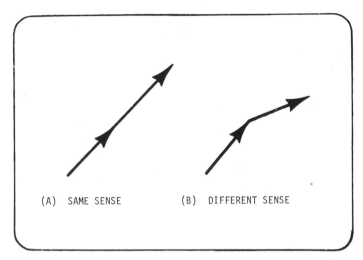

Figure 3.4   Combination of display vectors.

```
C     ***************************************************************
C     *                                                             *
C     * DISPLAY PROGRAM FOR A SYSTEM OF COPLANAR VECTORS.   *
C     * MAGNITUDES ARE IN KILOGRAMS AT THE CONSTRUCTION OF  *
C     * 10 DISPLAY UNITS EQUAL 1 KILOGRAM.                  *
C     *                                                             *
C     ***************************************************************
      CALL BEGIN
C     DISPLAY VECTORS 1,2,3,4 WITH ARROWHEADS
      CALL AROHD(1.,4.,2.,5.,.125,0.,16)
      CALL AROHD(2.,5.,4.,6.,.125,0.,16)
      CALL AROHD(4.,6.,5.5,4.5,.125,0.,16)
      CALL AROHD(5.5,4.5,4.,3.,.125,0.,16)
C     PROVIDES INTERNAL DATABASE
      REAL ICHAR
      DIMENSION X(8),Y(8),HT(8),ICHAR(8),ROT(8),NCARS(8)
      DATA X/ 1.25,2.5,5.,4.75,1.5,2.75,4.25,4.25/
      DATA Y/ 4.5,5.5,5.25,3.5,4.1,5.1,5.1,3.5/
      DATA HT/ 8*  .125/
      DATA ICHAR/ '1','2','3','4','150k','225k','212k','225k'/
      DATA ROT/4*0.,45.,30.,270.,60./
      DATA NCHARS/4*1,4*4/
C     PLOTS ANNOTATION ON DISPLAY
      DO 100 I = 1,8
  100 CALL SYMBOL(X(I),Y(I),HT(I),ICHAR(I),ROT(I),NCHARS(I))
      CALL FINITT(0, 0)
      STOP
      END
```

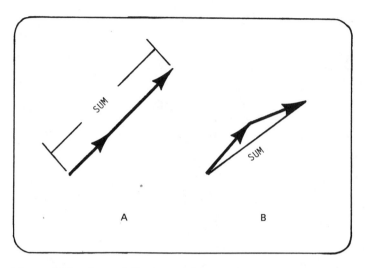

Figure 3.5   Sum of display vectors.

This display program was used to create Figure 3.6. However, a single reference to a subroutine;

CALL VECTOR(XARRAY,YARRAY,NV)

could have been used to replace the entire graphic portion of the display program. The vector subroutine was developed for any number of vectors to be displayed on the screen. It is as follows:

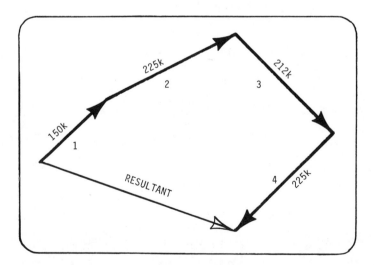

Figure 3.6   Resultant of multiple display vectors.

```
      SUBROUTINE  VECTOR(XARRAY,YARRAY,NV)
      D0 1000 I = 1,NV
1000  CALL AROHD(XARRAY(I),YARRAY(I),XARRAY(I+1),YARRAY(I+1),
      +.125, 0. , 16)
      RETURN
      END
```

The XARRAY and YARRAY need to be dimensioned in the display program
as NV+1, where NV stands for the number of vectors to be displayed.
The arrays can be loaded in the display program as

```
      DATA XARRAY/1.,2.,4.,5.5,4./
      DATA YARRAY/4.,5.,6.,4.5,3./
```

and the display created by the data statements is called a vector polygon.

Scales

In the CAMD solution of problems related to velocity, it is necessary to
display the machine element full size, smaller, or larger scale. This
display scale is expressed in three ways: (1) proportionate size (element
size multiplied by a factor, e.g., 2.,.6,1.5), (2) the number of display
positions on the CRT equal to a metric unit on the machine element, and
(3) a metric unit (e.g., 1mm) equals so many units on the machine
element. The display scale is designated S(k).
      The velocity scale, designated S(kv), is defined as the linear
velocity in distance units per unit of time represented by 1 display unit
on the CRT. If the linear velocity of a point is 5 mm/sec and the S(kv)
is 5, a line 1 unit long on the CRT would represent a linear velocity of
5 mm/sec and would be written

      $S(kv) = 5$

The acceleration scale, designated S(ka), is defined as the linear acceleration in distance units per unit of time per unit of time, represented by
1 display unit on the CRT. If the linear acceleration of a point is
100 mm/sec^2 and S(ka) is 100, a line 1 display unit long would represent
a linear acceleration of 100 mm/sec^2 and would be written

      $S(ka) = 100$

Resolution and Composition

      As described in the display vector section, the process of obtaining
the resultant (sum) of any number of vectors is called vector composition.
The reverse process of breaking up a single vector into two components is
called vector resolution.
      A vector resolution may be shown on a CRT by displaying two com-
ponents parallel to scalar lines making any desired angle with each other.
In other words, the sum or original vector will be the diagonal of a

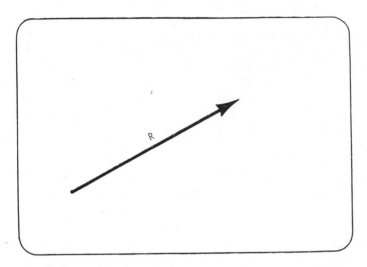

Figure 3.7   Resultant vector.

parallelogram obtained with the resolved vectors forming two of the
sides. The resolution process begins by displaying a resultant vector
called R, as shown in Figure 3.7. Vector R is composed of vectors A
and B, as shown in Figure 3.8. Vectors A and B can be resolved so that
a vector system can be displayed as in Figure 3.9.

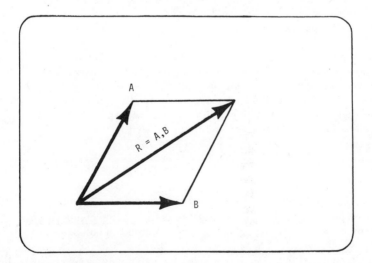

Figure 3.8   Components A and B.

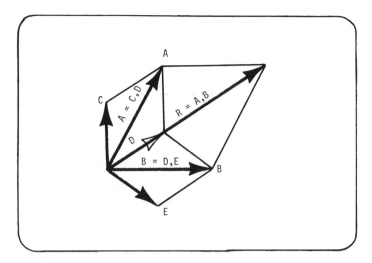

Figure 3.9   System of components

Obviously, this system is for illustration purposes, because the designer selected parallelograms at will. Resolution is a valid technique for the analysis of velocity polygons. For example, if the velocity of one point and the direction of the velocity of any other point on a machine element are known, the velocity of any other point on that element may be obtained by resolving the known velocity vector into components "along" and "perpendicular" to the scalar line joining these points on the CRT display.

Begin the process by displaying the two known points on the machine member, shown as A and B in Figure 3.10. The velocity of point A can be displayed as a vector, V(A), in Figure 3.11. The direction of the velocity of point B is known and can be displayed as a scalar line, called L in Figure 3.12.

Since this is a machine element, the distance between A and B is constant and a scalar line can be displayed through them as shown in Figure 3.13. Use the construction scalar to resolve V(A) into components along and perpendicular, as shown in Figure 3.14. Use the vector labeled "ALONG" in Figure 3.14 to extend the construction scalar display of AB and drop a perpendicular vector display at the tip of "along" to meet the direction line L. This is the perpendicular component of the velocity of point B, and the meeting of this component with the direction line L is the total velocity of point B, labeled V(B) in Figure 3.15.

In a more useful sense, suppose that a machine member has three points, labeled A, B, and C. The velocity of A is known, the direction of

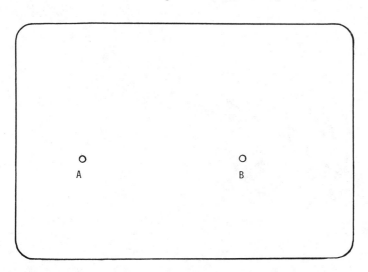

Figure 3.10   Display of two known points.

B is known, but neither the magnitude nor the sense of point C is known.
Using the techniques of resolution and composition, the total velocity of
point C can be found.   Begin the CAMD process by displaying the physical
location of the three points as displayed in Figure 3.16.   Display the

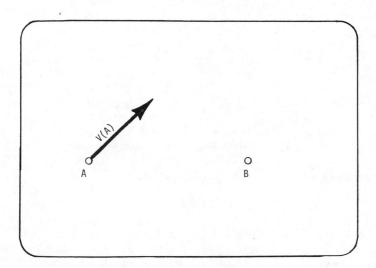

Figure 3.11   Display vector A.

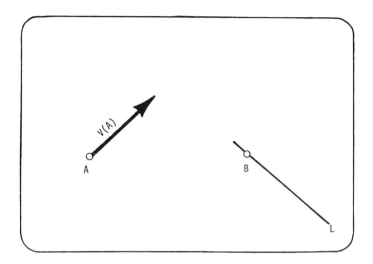

Figure 3.12   Sense of display vector B.

velocity of point A shown in Figure 3.17. Display the direction of point B
in Figure 3.18. Display the constant distance scalars through A, B, and
C as in Figure 3.19. Resolve V(A) into components along and perpendicular to scalar AB (Figure 3.20). Display the "along" component at

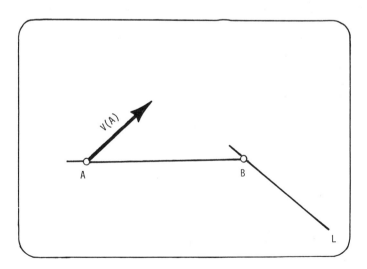

Figure 3.13   Display of scalar line AB.

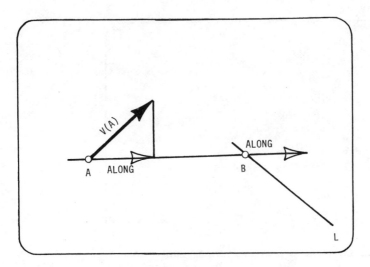

Figure 3.14   Resolution of V(A).

point B and drop a scalar connector to line L as indicated in Figure 3.21.
Resolve V(A) into components along and perpendicular to scalar AC, as
shown in Figure 3.22. Display the "along" component at point C and drop
a scalar connector called scalar M, as illustrated in Figure 3.23.
Resolve V(B) into components along and perpendicular to scalar CB

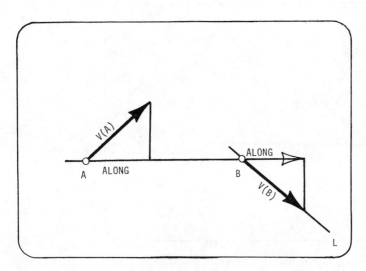

Figure 3.15   Composition of V(B).

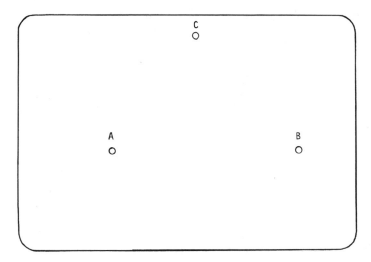

Figure 3.16   Display of three known points.

(Figure 3.24). Display the "along" component at point C and drop a scalar connector called scalar N, illustrated in Figure 3.25. This is the total velocity of point C, labeled V(C).

In this interactive velocity analysis, points A, B, and C were not located on a common line. In cases where three points are in a straight

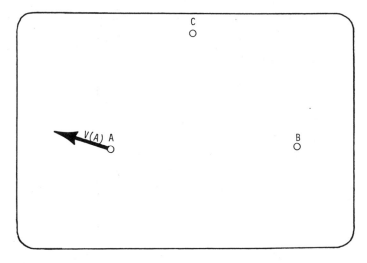

Figure 3.17   Velocity of point A.

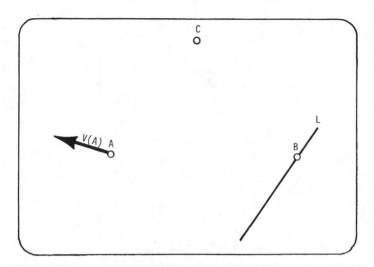

Figure 3.18   Sense of point B.

line, a ready solution for the velocities of B and C may be displayed when
it is realized that scalar L has angular motion about an axis of rotation
and that the velocity components perpendicular to B and A are proportional to each other.

   Begin the CAMD analysis by displaying the location of points A, B,
and C shown in Figure 3.26. Next, display the sense of point B and the

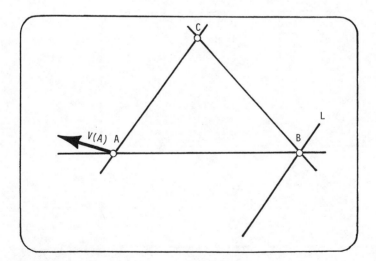

Figure 3.19   Scalar lines A B C.

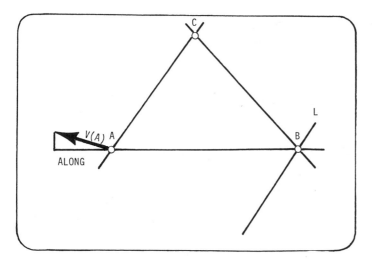

Figure 3.20   Resolution of V(A).

velocity of point A (Figure 3.26). Resolve the components of V(A) and display the "along" component to point B in Figure 3.27. Display a scalar AB and drop a scalar connector from the "along" component to scalar AB in Figure 3.28. This is the total velocity of point C.

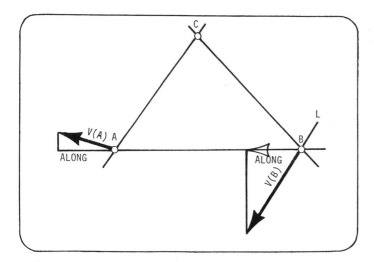

Figure 3.21   Composition of V(B).

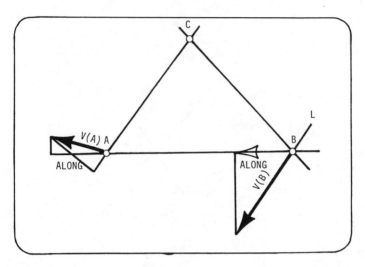

Figure 3.22   Resolution of V(A).

Axis of Velocity

Each member of a machine is rotating about either a fixed axis or a
moving axis. For display on a CRT, the moving axis may be frozen
for an instant at a time and treated as a stationary axis with display
properties similar to a fixed axis. This is the second method of dis-

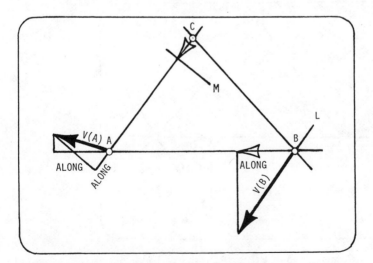

Figure 3.23   Composition of M.

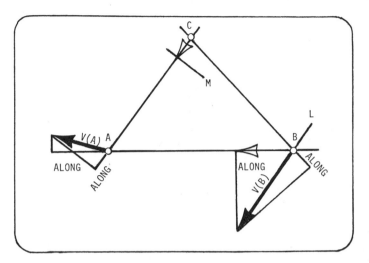

Figure 3.24 Resolution of V(B).

playing vectors and analyzing velocities and is considered to be as economical as resolution and composition.

Floating Links. The cranks of a machine rotate or oscillate about their respective fixed axes and the connecting elements, called floating links, rotate with an "absolute" angular velocity about an instantaneous axis of

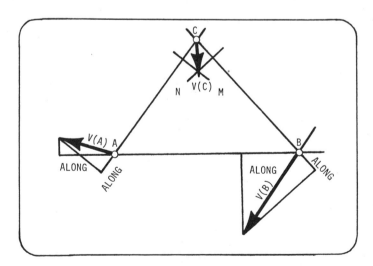

Figure 3.25 Composition of N.

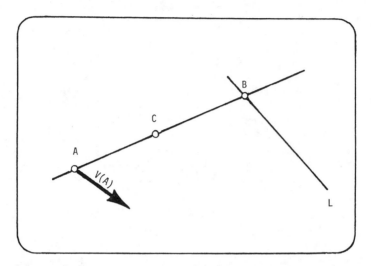

Figure 3.26   Display of points and sense of B.

velocity. The absolute instantaneous linear velocities of points on the
link are proportional to the distance of the points from the instantaneous
axis and are perpendicular to scalars joining the points with the instantaneous axis. Figure 3.29 represents a particular-shaped crank turning
about the fixed axis FA with an instantaneous angular velocity IA, producing the linear velocity of A represented by a display of V(A). The magni-

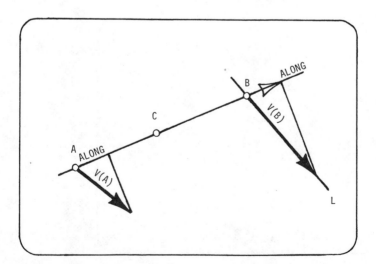

Figure 3.27   Resolution of V(A), composition of V(B).

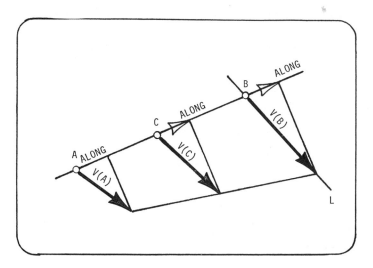

Figure 3.28   Scalars AB, tip connections, and compositions.

tude of the velocities of B and C are proportional to V(A) as their respective
distance from FA. The magnitudes are obtained by the use of similar triangles, not unlike the display method used for resolution and composition
(method 1). In each case the direction is perpendicular to AFA, BFA, and
CFA; and the sense of each linear velocity is consistent with the clockwise
angular velocity of the crank.

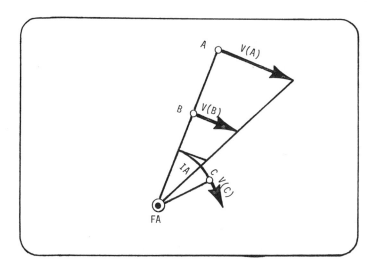

Figure 3.29   Axis of velocity.

Figure 3.30 represents a particular-shaped floating link. The absolute linear velocity of A is known and point B has a velocity sense of L. The instantaneous axis of velocity IV may be displayed by locating the intersection of the scalars perpendicular to the directions of the velocities of A and B. At the instant in time under consideration all points in the link are tending to rotate about IV. The vector V(B), displaying magnitude, can be positioned along the sense L by the use of similar triangles, as shown in Figure 3.29.

Rolling Bodies.   If a wheel, as displayed in Figure 3.31, rolls along the surface of H without slipping, the point of contact (IV) of the wheel and the surface is the instantaneous axis of velocity. The entire wheel acts as a crank rotating about the axis IV. The magnitudes of the velocities of points on the wheel are proportional to their respective distances from IV and are perpendicular to scalars joining the points with IV. If V(A) represents the velocity of A, the center of the wheel, then by similar triangles V(B) represents the velocity of B.

Instantaneous Axis.   It should be clearly understood at this point that (1) there is one instantaneous axis of velocity for each member in a machine, (2) there is not one common instantaneous axis of velocity for all elements in a machine, and (3) the instantaneous axis of velocity changes position as the element moves. The instantaneous axis of velocity can be displayed on a CRT whenever the directions of the velocities of two points on the element are known.

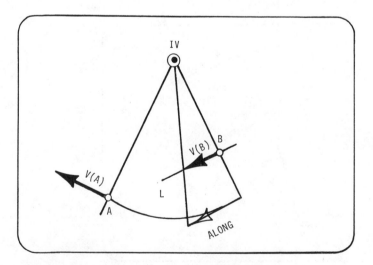

Figure 3.30   Instantaneous axis of velocity.

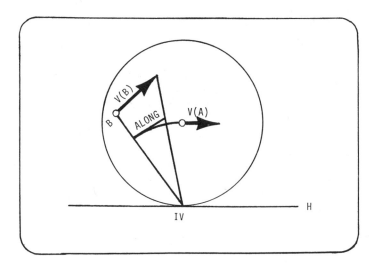

Figure 3.31   Instantaneous axis of rolling bodies.

The instantaneous axis of velocity is not an instantaneous axis of acceleration. Acceleration display methods are discussed in detail in Chapter 4. The instantaneous axis of velocity is a moving axis but may be displayed in a fixed position on a direct-view storage tube CRT. An instantaneous axis of velocity may have an acceleration; it does not necessarily equal zero acceleration as is the case for a fixed center of rotation. To display velocity and acceleration, a dynamic CRT display is required.

CENTROS IN COMPUTER-AIDED ANALYSIS OF MACHINE ELEMENTS

As previously stated when discussing display vectors, the instantaneous-axis-of-velocity method of display is a simplified version of the centro display method. In many mechanism displays, the instantaneous axis of rotation is not obvious, because the directions of the motion of two points on the element may not be apparent. By using the method of a centros, velocities in all machine mechanisms can be obtained.

Notation, Number, and Location

A centro may be displayed as (1) a point common to two elements displaying the same velocity, (2) a point in one element about which another element actually turns, or (3) a point in one element about which another element tends to turn. The last description of a display method is also a type of instantaneous axis display. It should be noted that a centro by the second method is permanently fixed and would be a location in the frame of a machine about which a crank turns. This second display would be a storage tube output, not a dynamic display. By the first method a point may actually be in two elements or a point in virtual space not actually in either element but assumed to be common in both elements. A display of this type should be refreshed in a dynamic mode.

All links, cranks, and elements (including the frame) are numbered on the computer display screen. Single notation is used (e.g., 1, 2, 3). The centro has a double-number notation, such as 12, 13, 23, 34, and so forth. The centro 34 (read "three-four") is in both machine elements 3 and 4. The total number of centros in a mechanism is the number of possible combinations of the elements taken two at a time. It may be obtained from the FORTRAN expression

$$NC = N*(N-1)/2$$

where NC is the number of centros and N is the number of elements in the displayed mechanism.

Centros are displayed on the DVST by a point-sorting routine which locates any three elements having plane motion relative to each other and sharing the same straight line. In other words, the three centros that are sorted and matched to each other lie "along" the same straight scalar display line. The meaning of "sorted" and "matched" should be further explained. Assume a four-link mechanism with the links numbered 1, 2, 3, and 4. From the FORTRAN expression it is determined that there are six centros: 12, 13, 14, 23, 24, and 34. Centros 12, 13, and 23 are sorted because if the common number in any two is cancelled, the numbers remaining will be the match of the third centro. Centros 14, 34, and 13 are sorted similarly, as are 24, 23, 34, and 14, 12, 24. By the use of the sorting subroutine, each of these four sets of matched centros lies on a straight scalar display. The total number of sets of matched centros depend upon the number of elements in the mechanism. In Figure 3.32, link 1 is the frame of the machine, 2 and 4 are cranks, and 3 is the connecting rod. The number and sets of the centros may be obtained from the sorting subroutine as

$$NC=4*(4-1)/2$$

or 6. The links were input as 1, 2, 3, and 4, while the matched sets were displayed as

12   23   34
13   24
14

Centros 12 and 14 are points in the frame (scalar 1) about which cranks 2 and 4 turn. Centro 23 is the geometric center of the pair 2 and 3; and 34 is for pair 3 and 4. The remaining centros cannot be found by the simple

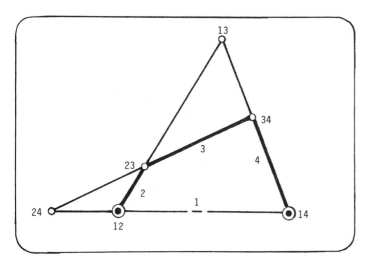

Figure 3.32   CAMD display of centros.

display of the mechanism. To locate centros 13 and 24, <u>interactive centro construction</u> is used. Again in Figure 3.32, centros 12, 23 and 34, and 14 were apparent from the display. However, the sorting routine indicated that two more centros were present (13 and 24).

Whenever a centro is located by simple display, interactively connect a scalar (by the use of a graphics tablet) so that a triangle can be formed. This new centro, whose number is the same as the numbers of the points joined, can be matched to the computer printout from the sorting routine. A scalar joining 2 and 4 locates centro 24; similarly, scalars from 23 and 34 locate centro 13. Additional examples of computer matching are given in Chapter 9.

Velocities of Centros

After the centros are located in Figure 3.32, the <u>linear</u> or <u>angular velocity</u> may be found. Using the same display as Figure 3.32, Figure 3.33 illus-illustrates the linear velocity of centro 23, 34, and 24. Since, by defini-tion, 24 is common to both 2 and 4 and has the same velocity, the deter-mination of the velocity of 24 would be

$$V(24) = V(23)$$

Of course, the velocity of the point 23 in link 2 must be known. By the use of similar triangles, the velocity of point 34 in link 4 can be found as

$$V(34) = V(24)$$

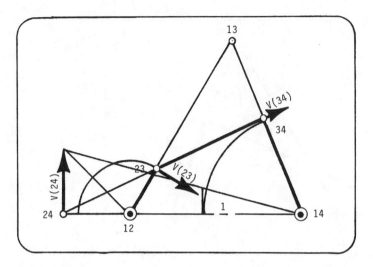

Figure 3.33   Velocities of centros.

The method of centros affords an excellent computer display for illustrating the angular velocity of any two links (e.g., 2 and 4):

V(24) = OMEGA(2)

and when 24 is considered to be in link 4:

V(24) = OMEGA(4)

Remember that the velocity of 24 is the same in each link:

OMEGA(4) = OMEGA(2)

The sense of rotation is displayed by labeling (notation) the desired link's angular velocity, corresponding to the sense of the linear velocity shown in Figure 3.33.

SUMMARY

This chapter described the technique of displaying (hardware and software) velocity analysis for the design of machine elements. Computer-aided machine design may be accomplished by strictly analytical means. This approach requires enormous amounts of software and in-house programming staff. An analysis by hardware/firmware is more direct, less complicated, and usually sufficiently accurate. This is achieved by the following:

Simplified software. This supports a display file maintaining separate
code sequences for each individually identified refreshed picture.
Any graphic output request that is not identified with a refresh
code is directed between refresh cycles to the storage display.

Graphic output. Requests to the system are made with a code plus
the X-Y location of the CRT. The code specifies the type of
graphics (e.g., MOVE, DRAW, DASH, DOT). To display a
rectangle in storage with sides of 50 display units by 100 display units, starting at CRT position 0.-0.; use:

Code	X	Y
MOVE	0	0
DRAW	50	0
DRAW	0	100
DRAW	-50	0
DRAW	0	-100

The CAMD system takes these commands and builds display code. The
interrupt service routine outputs the storage graphics between refresh
cycles.

Refresh display. The sequence of vector outputs listed above may
be saved and generated as a refreshed object. This is done by
preceding the vector output with an open command. This
builds the display code into dynamic memory areas. A close
command stops the display code.

Dynamic displays. Once a display file is built, several manipulation
commands can be used to affect its viewing status. A post
command links the display list for an object to the refresh display list and makes it visible on the CRT. Another command
will remove the list. Rapid interchange of these commands
will cause the object to "blink." A command may be used to
move an object from refresh to storage display, or another
can release the initial starting locations (X-Y) of an object to
display program control and the object will move (animation)
on the CRT.

Vector displays. An exchange status command initiates the sequence,
which allows the display controller to access the display list for
(1) resolution and composition, (2) instantaneous axis, and
(3) centros.

BIBLIOGRAPHY

Anderson, R. H., Storage cathode-ray tubes and circuits, Tektronix, Inc.,
    Beaverton, Oregon, 1968.
Cheek, T. B., Improving the performance of DVST display systems,
    Proceedings of the Society for Information Display, International
    Symposium Digest Technical Papers, April 1975.
Koenigsberg, L. K., A graphics operating system, Computer Graphics,
    Vol. 9, No. 1, Spring 1975.
Lauer, D. J., CAD interactive graphics systems designed by users,
    Proceedings of the 15th Numerical Control Society, Chicago, 1978.
Machover, C. W., Graphic displays, IEEE Spectrum, Vol. 14, No. 8, 1977.
Newman, W., and R. F. Sproull, An approach to graphic system design,
    IEEE Spectrum, Vol. 62, No. 4, April 1974.
Ryan, D. L., Computer-Aided Graphics and Design. Marcel Dekker,
    New York, 1979.
Shubert, R. A., NC interactive graphic systems, Society of Manufacturing
    Engineers Technical Paper No. MS77-971, given December 30, 1977.
Voelcher, H. B., Geometric modeling of mechanical parts and processes,
    Computer, Vol. 10, No. 12, 1977.

4

Interactive Acceleration Analysis

With the advent of extremely high speed computers (1970s) and the
associated engineering graphics hardware available today, the machinery
designer can save time and money. The designer is smart, creative, and
slow; whereas the computer and its various graphical displays are stupid,
uncreative, and very fast. The situation is thus a working relationship
whereby the person and the machine can interact. Certainly, their
characteristics complement each other, but their languages are very
different. Human beings think in symbols and pictures, whereas the
computer understands only simple electrical impulses. Hardware innovations such as the graphics tablet; whereby the designer may enter a
sketch for the computer to store as numerical data based upon a graphical
shape (database), now make possible interaction at a design level. This
interaction is a way of translating human instructions into electronic data,
and conversely, to convert the computer's impulses into engineering
documents.

ACCELERATION IN MACHINE DESIGN

Interactive computer display techniques have made the study and analysis
of the accelerations in moving machinery extremely important. An inertia
produced by the acceleration of the parts of a machine may be studied at a
high magnitude. In some situations, at a certain position in a cycle, the
forces displayed on the designer's console may be the only visual study
ever produced by the working design. This new technique for analyzing
the acceleration of points in the links of a design should be the prerequiste
for the inertia force analysis of the desired machine design. In this
chapter the display technique for the acceleration in machine members
is presented. The method developed, is similar to the computer-aided
velocity analysis for cathode ray tube (CRT) display presented in Chapter 3.

Point Acceleration

The acceleration of a display point is the time rate of change of the velocity
of the displayed point. Since computerized display of velocity (Chapter 3)
has both magnitude and sense, there may be a change in the velocity
magnitude, sense, or both. The rate of change in the display velocity
in sense is the normal component of the resultant acceleration of the
display point. This is called normal acceleration.

Curved Paths.   The rate of change in the display velocity in magnitude is
the tangential component of the resultant acceleration of the display point
and is called tangential acceleration. In Figure 4.1 the display point P is
in animation (motion) along the curved path anotated L with radius R at
center FA. The angular velocity of the component P(I)FA is ω (OMEGA).
The magnitude of the linear velocity of P(I) about FA, called V(P(I)FA), is
OMEGA*R and is the perpendicular component of FAP(I). At the end of a
time interval, P has reached P(I). PFA has turned through animation and
is now represented by P(I)FA. The angular velocity of PFA is therefore
OMEGA + ΔOMEGA. The linear velocity of display point P about FAV(P,
FA) is (OMEGA + Δ OMEGA)*R and is the perpendicular component of
FAP.
        On the graphics tablet shown in Figure 4.2 the velocity polygon can
now be constructed as follows:

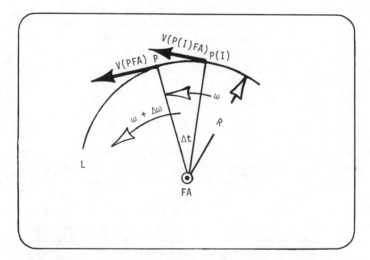

Figure 4.1   Acceleration of a display point in curvature.

Figure 4.2 Interactive graphics tablet. (Courtesy of Tektronix, Inc.)

1. Begin the construction with FAP(I) equal to V(P,FA) in a convenient portion of the display screen.
2. Then set FAP equal to V(P,FA).
3. Now P(I)P is connected, illustrating the velocity of P relative to P(I), and is the increase in the velocity of P(I) in the time interval Δt.
4. With the aid of an ordinary compass and the electric pencil of the graphics tablet, interactive acceleration analysis can now be used.
5. Set the legs of the compass equal to FAP(I) and construct an arc through PFA. Label this intersection P(a).
6. Between points P(I) and P(a), call for a display vector from computer memory (Chapter 3).
7. The change in the linear velocity of P is equal to the vector sum of P(I)P(a) and P(a)P.
8. The vector P(I)P(a) is the normal acceleration of P. The tangential acceleration of P about FA is represented by the vector P(a)P.

The foregoing procedure, shown in Figure 4.3, leads to the following observations.

1. When a display point is in animation about an axis, the display point has a normal acceleration about the axis, regardless of whether the scalar display joining the display point and the axis has an angular acceleration.
2. The magnitude of this normal acceleration is related to the angular velocity of the scalar display joining the two points and the distance between them.
3. Whenever a display point is in animation relative to another (turning), the first display point has a tangential acceleration about the second display point, provided that the joining scalar has angular acceleration.
4. The sense of the tangential acceleration is perpendicular to the joining scalar.
5. The magnitude of the tangential acceleration is related to the angular acceleration of the joining scalar and the distance between them.
6. The resultant linear acceleration of point 1 about point 2 is the vector sum of the normal and tangential accelerations.

Two-Point Linkage.   In the display of acceleration (Figure 4.1), no reference was made as to whether FA was fixed absolutely or relatively. The observations discussed are correct for both. However, in this section a method is developed for obtaining the absolute linear acceleration

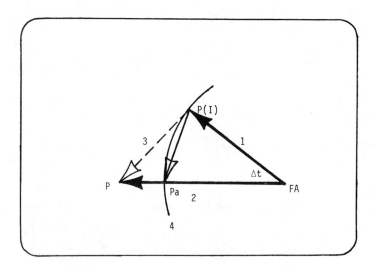

Figure 4.3   Construction of a velocity polygon.

of a point on a machine element when the absolute linear acceleration of
another point on the element is known. In Figure 4.4, link 1 is turning
about FA with an angular velocity of $\omega(1)$ and an angular acceleration of
$\alpha(1)$. The floating link 2 is pinned to P(1) and is turning with an angular
velocity of $\omega(2)$ and an angular acceleration of $\alpha(2)$. At this Δt the
linear velocity of P(1) is V(P(1)) and the linear velocity of P(2) is V(P2)).
At the end of Δt, P(1) has moved to P(1,I) and P(2) has moved to P(2,I),
as shown to the left in Figure 4.4. FA remains fixed while the velocities
of P(1) and P(2) are shown at the end of Δt. Using the interactive acceleration analysis technique, enter the velocities of each position as shown
in Figure 4.5. FA remains fixed, but the display origin is moved and
plotted for better visibility in Figure 4.4. The velocities for each display
point are shown in the newly created display position, while the changes in
the velocities are represented by $\Delta\alpha$ P1$_i$,P2$_i$ and $\Delta\alpha$ P1,P2 in Figure 4.5.

Relative Methods. The principles of interactive acceleration analysis
will provide a visual display method of obtaining the linear accelerations
of points in the links and the angular accelerations of the links in a machine.
Study the display of the machine element (ME) section shown in Figure 4.6.
The absolute acceleration of display point A is indicated as A(IA), the
velocity of A is displayed as VA, the sense of the velocity of display point
B is along scalar BL, and the direction of the acceleration of B is
assumed to be along BO.
    The linear acceleration of B and the angular acceleration of AB can
be found by the following interactive construction techniques:

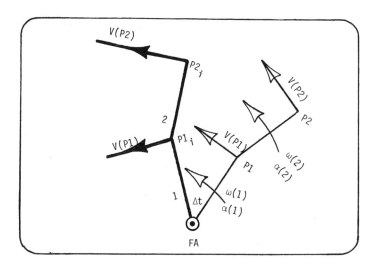

Figure 4.4 ΔT display of two-point linkage.

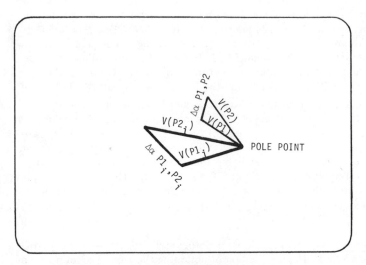

Figure 4.5   Interactive acceleration analysis.

1. Select an unused portion of the face of the CRT to display velocity vector VA as shown in Figure 4.7.
2. Display the sense of the velocity of point B from the tail of VA.
3. Display the sense of AB through the tip of vector VA as in Figure 4.8.
4. Display a perpendicular connector from the scalar AB intersecting the sense of B.

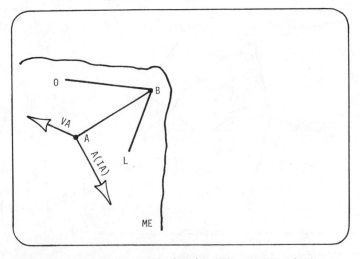

Figure 4.6   Relative methods of acceleration analysis.

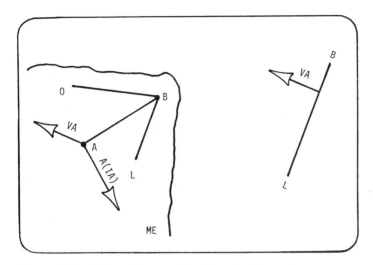

Figure 4.7   Interactive construction technique.

Now the acceleration can be displayed. Absolute resultant accelerations originate at the pole point selected by the designer, while relative
accelerations originate at the termini of the absolute accelerations. From
the pole point, display a vector equal in length and parallel to A(IA) shown
in Figure 4.8. From the tip of A(IA), display a scalar representing the
normal acceleration of B about A, parallel to BA. In Figure 4.9 the sense
from B to A is equal to D in the interactive steps shown. Next, display

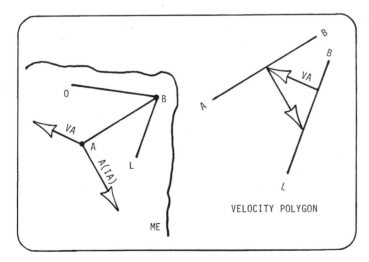

Figure 4.8   Steps 3 and 4 of construction.

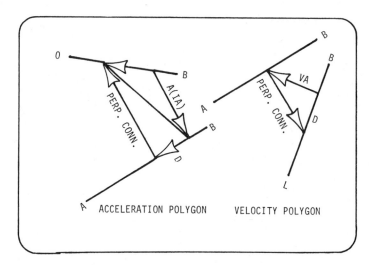

Figure 4.9   Acceleration construction diagram.

a scalar representing BO through the pole point. A perpendicular construction vector from the tip of D intersection BO and a construction scalar from
this intersection to the tip of A(LA) complete the acceleration diagram.

Display of Normal Acceleration by Plotter

The normal acceleration of a point about another point may be obtained
graphically as plotter output when a plotter scale relationship is applied.
In Chapter 3 the display scales for CRTs were defined. Since the display
scale was defined three ways, one definition only may be used for plotter
output of normal acceleration construction.
        The scales used for plotter output are as follows:

S(K) = the physical size of the machine part that 1 cm on the plotter
represents; e.g., if the plotted image is one-half size, S(K) = .647

S(IV) = the velocity in units per second that 1 cm on the plotter
represents

S(IA) = the acceleration in units per second that 1 cm on the plotter
will represent

        In Figure 4.10 a machine element labeled FA,A was plotted on a
drum plotter to the S(K) scale and represents counterclockwise turning
about FA with an angular velocity that produces a linear velocity of A,
represented to the S(IV) scale by vector AM. Therefore,

        V(FA,A) = AM*S(IV)

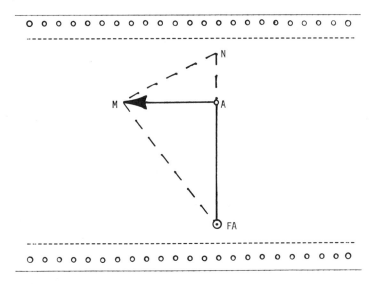

Figure 4.10   Plotter output of acceleration.

while FA,M is plotted with the command

CALL DASH(FA(X,Y),M(X,Y),DELTA)

where FA(X,Y) are the current plotter coordinates for FA and M(X,Y)
are for plotter location M. DELTA represents the length of each dash
segment. The dashed segment NA is the plotted representation of the
normal acceleration of A about FA to the S(IA) scale. The scales have
been chosen so that

$$S(IA) = S(IV)**2/S(K)$$

and therefore triangles MNA and FA,MA are similar. AN represents, to
the S(IA) scale, the magnitude of the normal acceleration of A relative
to FA.

Acceleration Polygons.   To understand the display of normal acceleration by plotter, the acceleration in a machine mechanism called a crank
and rocker is shown plotted in Figure 4.11. This example is the same as
the relative method except that the normal accelerations are plotted on an
X-Y plotter.

Since plotter output is used, the scale relationship

$$S(IA) = S(IV)**2/S(K)$$

must be observed. Crank 2 has an angular velocity of 200 rpm counterclockwise and a negative angular acceleration of 280 rad/sec^2. For the

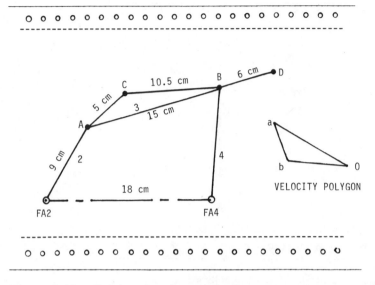

Figure 4.11   Crank and rocker mechanism.

position plotted, the instantaneous linear accelerations of A, B, C, and D and the instantaneous angular velocities of links 3 and 4 can also be plotted by the use of acceleration and velocity polygons.

<u>Velocity Polygons</u>.   The display of the velocity and acceleration polygon for the crank and rocker is as follows;

     1. Rerun the plotter program and plot the information contained in Figure 4.11. Modify the output program to calculate VA and plot it perpendicular to FA2,A. Change the color of the plotter lines by

     CALL NEWPEN(3)

Now write and enter the commands to connect points M and N as shown in Figure 4.10. This will plot the first acceleration polygon, as shown in Figure 4.12.

     2. Repeat the process for FA4,B and create an output acceleration polygon for VB as indicated in Figure 4.12. The second acceleration polygon is now sent to the plotter and can be viewed as shown in Figure 4.2 of the graphics tablet and interactive plotter.*

*The complete working description of interactive plotters can be found in Ryan (1979).

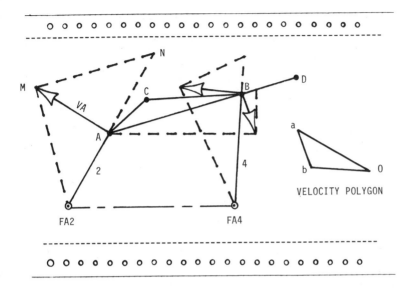

Figure 4.12   Rerun of crank and rocker.

3. Create the acceleration polygon for VBA following the steps shown
    in Figure 4.10 and plotted in Figure 4.12. The third acceleration
    polygon is now sent to the plotter.
4. The velocity polygon can now be created by the construction steps
    shown in Figure 4.3.

Velocity polygons are often displayed as plotter output for mechanisms
containing cranks and slides. The slider-crank mechanism probably is used
in more machines than any other mechanism. For this reason special
techniques are given for the output of velocities and accelerations in this
mechanism. In Figure 4.13 a slider-crank mechanism is plotted to the
S(K) scale.

Display of Klein's Method

Special solutions for obtaining the velocities and accelerations in a slider-
crank mechanism have been developed. One of these techniques, known as
Klein's construction, is shown in Figure 4.14. At FA erect a perpendicular
to the path of travel of the slide. At the intersection of this perpendicular
construction with the connecting rod, locate W. With AW as a radius and
A as a length, plot a circle. With the radius MA, equal to one-half the
length of the connecting rod (AB) and a center a M (midpoint of the rod),
plot a circle by

        CALL CIRCLE(A(X,Y),WA,0.,60,6.)
        CALL CIRCLE(M(X,Y),MA,0.,120,3.)

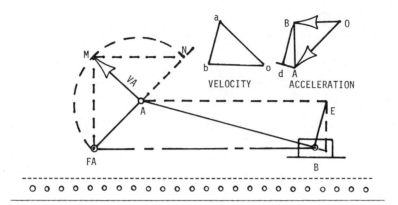

Figure 4.13   Display of slider-crank.

where A(X,Y) are the plotter coordinates for the center of the first circle
and M(X,Y) are the plotter coordinates for the second circle to be plotted.
WA and MA are the respective, radius for circle 1 and circle 2; the remaining constants are used to specify the smoothness of the plotted line.
     Now join the intersections T and J of these two circles by

        CALL PLOT(XT,YT,3)
        CALL DASHP(XJ,YJ,.01)

where XT and YT are the coordinates of a pen position and XJ and YJ are
the end points for a dashed line. As a check on the method, TJ will be

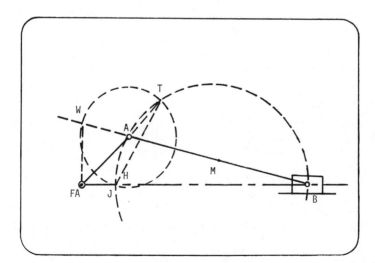

Figure 4.14   Display of Klien's methods.

perpendicular to AB. Position H at the intersection of TJ and a line
through FA parallel or coinciding with the path of travel of the slide by

    CALL PLOT(XFA, YFA, 3)
    CALL CENTER (XB,YB,TLEN,DASH,0.)

where XFA and YFA are the coordinates of the center line to be plotted and
XB and YB are the end points of the center line. TLEN is the distance
between FA and B and DASH is the size of the break in the center line
displayed in Figure 4.14. HFA is equal to the linear acceleration of the
slide (B) to the S(IA) scale.

Interactive Construction Techniques.   For interactive construction to be
applicable, the following conditions must be satisfied:

1. The mechanism must be a slider-crank mechanism with the
    guides stationary.
2. The crank must rotate at a uniform velocity to be output on a
    CRT.
3. The scale relationship S(IA) = S(IV)**2/S(K) must be used.
4. The length of the crank FA,A on the screen must be equal to
    the scale S(IV), the linear velocity of the crank A.
5. The length of the crank FA,A on the screen must be equal to
    the S(IA) scale, the normal acceleration of the crank A.

In Figure 4.14 the triangle FA,AW is the velocity polygon turned
through 90°. FA,A is equal and perpendicular to AO of the velocity
polygon of Figure 4.13. FA,W is perpendicular to BO and AW is
perpendicular to AB in the velocity polygon of Figure 4.13. Therefore,
triangles FA, AW, and OAB are equal.

The proof of Klien's construction is based upon its similarity to the
display of normal acceleration by plotter. While the graphical procedures
are well established for Klien's construction, the introduction of computer
graphics and the high speeds of presentation make this technique extremely
useful for slider-crank mechanisms.

Points on a Rolling Body.   The acceleration of any point on a body that
rolls without slipping on another machine surface is determined by the
proper application of the principles already described. In general, the
most convenient interactive method is as follows:

1. Find the acceleration of the center of curvature of that part of
    the body that is in contact with the machine surface.
2. Find the acceleration of the given point relative to the center
    of curvature.
3. Find the vector sum of these two quantities, which will display
    the required acceleration of the point in question.

Display of Coriolis Method

In Figure 4.13, the guide for the sliding member was fixed to the earth.
There are a number of machines where the guide as well as the slide moves
with respect to the earth. For this case the plotter method developed earlier
cannot be used, but the acceleration may be obtained by the use of Coriolis's
law* and displayed on a CRT.

> Coriolis's law: when a particle is moving along a path that is
> also in motion, the absolute linear acceleration of this particle
> is the vector sum of (1) the acceleration the particle would
> have if the path were fixed and the particle moved only in the
> path, (2) the acceleration the particle would have if the
> particle were fixed to the path and the path moved, and
> (3) a compound supplementary acceleration called Coriolis
> acceleration. The compound supplementary acceleration
> is equal to twice the product of the velocity of the particle
> relative to the path and the angular velocity of the path.

Stated in FORTRAN form, Coriolis's law is:

$$AP = ALP+ALFA+(2*U*W)$$

where AP is the absolute linear acceleration of the particle (point P) in
Figure 4.15. ALP is the linear acceleration of point P relative to the

*The theory of Coriolis's law is rather involved. No attempt will be
made to prove the law. For a complete discussion of this law, the
reader is referred to Paul (1979).

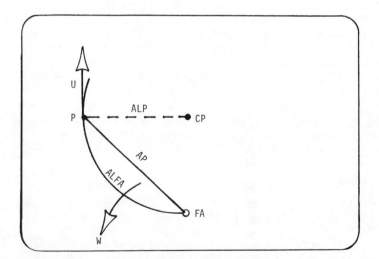

Figure 4.15   Display of the Coriolis method.

center point CP in Figure 4.15. ALFA is the absolute linear acceleration of point P about FA, and 2UW is the calculation for the compound supplementary acceleration. 2UW is a normal acceleration and is dependent upon the changes of the directions of the velocity of point P. A rule for the direction (sense) of 2UW is that it is always parallel to PCP and perpendicular to U.

There could be some confusion as to when to use the Coriolis method and when to use the relative acceleration technique discussed earlier in this chapter. Whenever a particle is moving on a body that is moving relative to the earth, use the Coriolis method. When displaying Coriolis's technique, the relative angular velocity and acceleration of the line joining the particle and the center of curvature must be used. When displaying the relative acceleration technique, the absolute angular velocity and acceleration of the line connecting the two points on the CRT must be used.

SUMMARY

Interactive computer display techniques have made the study and the analysis of the accelerations in moving machinery extremely easy to visualize. For example, the inertia produced by the acceleration of the parts of a machine may be displayed on the face of a CRT and studied at a high magnitude. In some situations the forces displayed on the designer's console may be the only visual study ever produced by the working design. This new method for displaying the acceleration of points in the elements of a machine should be the prerequisite to the inertial force analysis of the desired machine. In this chapter the display techniques for the acceleration were:

1. Point acceleration
2. Relative acceleration
3. Normal acceleration for plotter output
4. Klein's method
5. Coriolis method

The methods displayed were similar to the computer-aided velocity analysis for CRT presented in Chapter 3. Collectively, the acceleration display methods were presented as "interactive acceleration analysis."

APPENDIX — GRAPHICAL INTERACTIVE PROGRAMMING (GRIP)

GRIP is a registered trademark of United Computing Corporation, a subsidiary of McDonnell Douglas Corporation. GRIP is a FORTRAN-like language. The GRIP compiler processes English-oriented free-form source code and generates object programs which are stored in a library by the program name. Source programs can be input in batch form (from cards) or interactively through the interactive design station

(IDS). Entities, parameters, strings, labels, and arrays can be assigned symbolic names. Their values can be specified interactively at execution time.

With the IDS, the user is able to develop, manipulate, and modify designs on a rela-time basis. Design documentation and engineering drawings evolve as the user enters parameters and selects system functions in set steps. Through the power and flexibility of the IDS, changes and additions can be observed instantly. Critical areas can be examined in minute detail from any view attitude by the use of zoom and view control functions. Conversely, desk programming provides for the writing of macros through the use of symbolic nomenclature even before parameter values have been defined. Macros (previously programmed logic), which establish set procedures, can be called time and time again to test, branch, loop evaluate, and display operations.

GRIP combines interactive and code sheet programming to capture the advantages of both approaches in a single process. Designs can be grouped into families and their aggregate descriptions programmed parametrically. The subsequent entry of a few key values during program execution can cause the entire drawing for a member of the family to be automatically generated. This formation of a GRIP lubrary facilitates the retention and ready availability of optimized programs that define recurring construction sequences unique to user requirements. Program nesting enables the user to write macros with a minimum of code. The programs written for graphical display can be optimized to ensure uniformly high quality with the shortest possible development time.

A.1 GRIP Commands

The GRIP language contains five commands for program communication with the IDS operator-machine designer. The command formats enable the designer to output messages to the IDS operator via the IDS message monitor (menu display). The five types of responses are:

> Indicate screen position
> Key-in text
> Select screen entities
> Option list selection
> Numerical data entry

It is particularly interesting to note that most of the conceptual framework of GRIP can also be found in the CORE concept of the ACM/SIGGRAPH Graphics Standards Planning Committee, which GRIP predates.

A.2 GRIP Geometry

Geometric entities are developed, displayed on the IDS screen, and entered into the database by the execution of the GRIP program. Entities

may be displayed when defined or suppressed by the DRAW/OFF command.
GRIP provides the following geometry definitions:

Geometry	Syntax
Point	POINT/X,Y,Z
Line	LINE/X_1,Y_1,X_2,Y_2
Circle	CIRCLE/X,Y,Z,R,SANG,EANG
Fillet	FILLET/ENT_1,ENT_2,CENTER, POINT,RADIUS,NOTRIM
Ellipse	ELLIPS/POINT,MAJOR,MINOR, ATANGL,START,END
Hyperbola	HYPERB/POINT,HTAXIS,HCAXIS ATANGL,START,END
Parabola	PARABO/POINT,FOCALL,YMIN, YMAX,ATANGL
General conic	GCONIC/A,B,C,D,E,F
Cubic spline	SPLINE/MOVE,DIST,SANG,EANG, $POINT_1$,$POINT_2$,$POINT_3$

In addition to the geometric entities, drafting functions are provided
as part of the GRIP language structure. A machine designer may select
from the following items:

Text, labels, and ID symbols
Decimal, fraction, and tolerance formats
Character-size control
Line fonts and density control
View selection and specification
Entity grouping
Blanking and unblanking
Automatic crosshatching
Translation and rotation
Mirroring

GRIP provides a wide range of options for the dimensioning of drawings. A comprehensive vocabulary enables the specification of dimension
type, origin, orientation, text size, and associated tolerances. The types
of GRIP dimensioning are: (1) horizontal, (2) vertical, (3) perpendicular,
(4) parallel, (5) angular, (6) radial, and (7) ordinate.

A.3 GRIP Programs

The GRIP software is supplied with user documentation. The following program listed is an example of a GRIP procedure for an IBM 370
model 3033.

```
SYMBOL              'SCREEN'
POLYGON             'LINE/O,O,IDATA1'
DOTSET              'NODES/0,0,IDATA2'
INSTANCE            'C/C1,0.3,0.3'
INSTANCE            'C/C2,0.6,0.6'
INSTANCE            'R/R1,0,0.3'
INSTANCE            'R/R2,0.3,0.3'
TEXT                'RNAME/0.2,0.33,IDATA3'
TEXT                'RVALUE/0.2,0.22,IDATA4'
TEXT                'CNAME/0.17,0.1,IDATA5'
TEXT                'CVALUE/0.17,0.8,IDATA6'
END

SYMBOL              'R'
END

SYMBOL              'C'
END

SEGMENT             'DIAGAM/0,0,IV'
INSTANCE            'SCREEN/F1,0.5,0.9'
SHOW

END
```

The machine designer will notice that GRIP is very procedure-oriented. This fact makes the foregoing program straightforward. A line-by-line explanation of the program will relate the naming of parameters.

```
SYMBOL                  'SYMBOLNAME'
```

SCREEN identifies the symbol (subpicture) created by the statements following the SYMBOL statement. The items in the primitive structure are not individually identifiable.

```
POLYGON 'ITEMNAME/MX,MY,IDATA,IV'
```

ITEMNAME identifies the geometric form, a line in this case. MX and MY represent the item origin, relative to the segment origin (absolute if MXA, MYA and relative if MXR,MYR). The parameter IDATA is an array of coordinates relative to MX and MY. IV is the item characteristics: for example; IV may be B, L, I, or P, where B is the blink status, L the character size, I the boldness or color, and P the lightpen status.

```
DOTSET 'ITEMNAME/MX,MY,IDATA,IV'
```

The itemname is NODES in the third line of the program; its origin is zero for MX and zero for MY. IDATA is a DIMENSIONED parameter, where IDATA(3,N) has been set aside for storage. N = number of primitives in the DOTSET. If IV is not specified, the software assigns a default value of B = 0, L = 0, I = 0, and P = 0.

INSTANCE 'SYMBOLNAME/INSTANCENAME,MX,MY,IV'

In program lines 4 through 7 the SYMBOLNAME is C and R, and the INSTANCENAMES are C1, C2, R1, and R2. In this case the R and C symbols are to be activated twice for C and twice for R. Separate INSTANCE storage locations are created (C1 and C2 for C, while R has R1 and R2), having the relative origin and the placement specified by MX, MY, and IV.

TEXT 'ITEMNAME/MX,MY,MODE,MESAGE,IV'

The next four lines of the GRIP program provide labeling for the sub-picture R and C. In storage location RNAME a label may be placed by reading the contents of MESAGE. MODE is the character write mode of the display device used.

A symbol definition section of GRIP programming is formed by the delimiters SYMBOL and END. Symbols may consist of an arbitrary number of items or other symbols nested in other symbol definition sections, as is the case for the SYMBOL sections 'R' and 'C', which appear next. A SEGMENT definition section of a GRIP program begins with SEGMENT and ends with END. The SEGMENT 'diagam' contains the SYMBOL 'SCREEN' just defined and can now be displayed with the command SHOW. The last statement END completes the SEGMENT 'DIAGAM'.

BIBLIOGRAPHY

Anderson, R. H., Storage cathode-ray tubes and circuits, Tektronic, Inc.
    Beaverton, Ohio, 1968.
Boguslavsky, B. W., Elementary Computer Programming in FORTRAN.
    Reston Publishing, Reston, Va., 1974.
Chasen, S. H., Geometric Principles and Procedures for Computer
    Graphic Applications. Prentice-Hall, Englewood Cliffs, N.J., 1978.
Giloi, W. K., Interactive Computer Graphics. Prentice-Hall,
    Englewood Cliffs, N.J., 1978.
Levinson, I. J., Machine Design. Reston Publishing, Reston Va., 1978.
Newman, W., and R. F. Sproull, Principles of Interactive Computer
    Graphics. McGraw-Hill, New York, 1973.
Paul, B., Kinematics and Dynamics of Planar Machinery. Prentice-Hall,
    Englewood Cliffs, N.J., 1979.
Ryan, D. L., Computer-Aided Graphics and Design. Marcel Dekker,
    New York, 1979.

5

Computer-Aided Linkage Design

The difference in meaning of the terms "computer-aided machine design" and "computer-aided design of machine elements" is a difference in viewpoint. In computer-aided design of machine elements, we use the term mechanism when dealing with the relative positions and the connections of machine parts. Also, their relative velocities and accelerations are displayed for analysis without regard to the shape of the parts. In computer-assisted design of machine elements, a mechanism is a machine in which the forces transmitted are fairly small, such as a clock, timing mechanisms, measuring instruments, or scale models. In computer-aided machine design, a machine may contain a number of integrated mechanisms, such as the automobile engine. Therefore, when a machine part is displayed as a part of a mechanism, it is called a link.

FOUR-BAR LINKAGE DISPLAYS

In Chapter 1 a mechanism was defined as a combination of links so connected that relative motion of any two compels definite relative motion of every part of the mechanism, and the four-bar linkage was evolved for display on the face of a CRT. Chapter 2 treated motion with this definition to allow us to display combinations of machine parts that conform to the laws of mechanism and combinations that did not—so that mechanisms could be displayed as animations on the CRT. Chapters 3 and 4 presented computer-aided methods for obtaining the linear velocities and accelerations of the displayed points on links and the angular velocities and accelerations of the links in a mechanism. With these chapters as preliminary preparation, some of the more common computer-aided machine designs may now be presented; with the intent of using these techniques to study:

1. Computer-generated transmission paths
2. Gears and cams
3. Flexible connectors
4. Other machine elements

In programming the display motion of a mechanism by using the relationship of the four-bar linkage, the first step is always to identify the
four-bar linkage or chain of linkages. It must be remembered that each
scalar display representing a link is part of a rigid body. The scalar of
centers is on a body assumed to be fixed. The center lines of the cranks
are also on rigid bodies turning about axes attached to this fixed body, and
the center line of the coupler is on a rigid body connected to each crank by
either a turning pair or a sliding pair.

To display a single link on the face of the CRT, the designer will
enter the following:

```
//JOBNAME JOB (0921-1-410-00-X,:06,1),'RETURN TO MAILBOX 38'
//STEP1 EXEC FTG1CLG,TEKPLOT=1812
//C.SYSIN DD *
      CALL  INITT(240)
      CALL  LINK(0.,3.5,8.,3.5)
      CALL  FINITT(0,0)
      STOP
      END
      SUBROUTINE  LINK(XPAGE,YPAGE,EX,EY)
      CALL  CIRCLE(XPAGE,YPAGE,.05,0.,60,6.)
      CALL  PLOT(EX,EY,2)
      CALL  CIRCLE(EX,EY,.05,0.,60,6.)
      RETURN
      END
//G.SYSIN DD*
/*
//
```

The subroutine LINK uses two CRT-designed display entities, PLOT and
CIRCLE. Each link with begin and end with a small, open circle that can
be used to represent a pin joint, fixed axis, or other connection by the
addition of graphical entities such as a dot (closed circle). The display
of this single link is presented in Figure 5.1.

To annotate the links on a display screen, it is easiest to start at
the driver and find the fixed axis about which the driver turns. Label
this member "CRANK." This is done with the statement

CALL LABEL(XPAGE,YPAGE,CHT,'CRANK',DEG,NCHRS)

where XPAGE and YPAGE are the screen coordinates for the start of the
label and CHT is the desired height of the character string to be plotted.

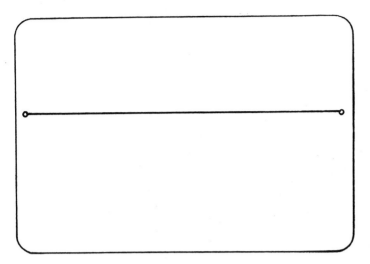

Figure 5.1   Display of single link.

DEG is the amount of rotation necessary to position the characters
"CRANK" parallel to the link displayed on the screen. NCHRS stands
for the number of characters in the label "CRANK," in this case, 5.
    Next label the fixed axis "FA" and determine the rigid piece turning
about FA, which is displayed by CALL PLOT. To display the fixed axis
on the face of the CRT, the designer will add the following to the display
program presented earlier.

```
      CALL INITT(240)
C     INSERT THE CALL TO DISPLAY A FIXED AXIS HERE
      CALL FA(1.,1.5,.25)
      CALL LINK(2.5,3.5,6.,3.5)
      CALL FINITT(0, 0)
      STOP
      END
      SUBROUTINE FA(XRB,YRB,DIARB)
      CALL DOT(XRB,YRB,DIARB)
      CALL CIRCLE(XRB,YRB,DIARB,0.,60,6.)
      RETURN
      END
```

The subroutine FA uses two CRT-designed display entities, DOT and
CIRCLE, as shown in Figure 5.2. Now the second crank can be displayed
by

```
      CALL PLOT (6.,3.5,3)
      CALL PLOT(6.,1.5,2)
      CALL FA(6.,1.5,.25)
```

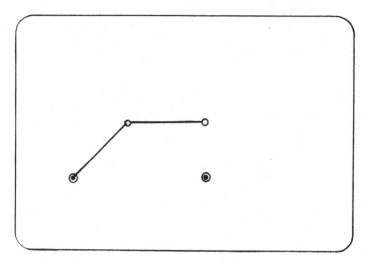

Figure 5.2   Display of FA and single link.

when these statements are inserted before the CALL FINITT statement.
This member can be labeled "CRANK 2." The member where the two fixed
axes are displayed is the fixed link, and the scalar joining the two fixed
axes can be called line of centers. In Figure 5.3 the driving crank trans-
mits motion to the driven crank through an intermediate connector displayed

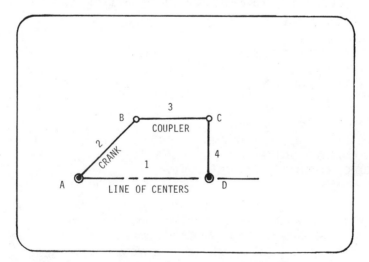

Figure 5.3   Display of completed four-bar.

by CA LL LINK. This link is labeled "COUPLER" in Figure 5.3. The
display scalar between the fixed axes is added to the display by

    CALL CENTER(1.,1.5,6.,1.5)

and all notation and labeling of Figure 5.3 can be done by

    CALL LABEL

The four-bar linkage having been displayed can now be properly labeled
and prepared for animation on the screen. It is not necessary that
linkages be displayed at rest. The display device is dynamic and can
present the relative motions of the four links. The fixed link shown in
Figure 5.3 may be attached to some other part of the machine which is
in motion. The relationship of the relative motions of the four links of
a four-bar linkage are unchanged, however. In other words the screen
location of the four-bar display can be in motion. The following program
will illustrate fixed link motion along the X axis in a positive direction:

```
      CALL INITT(240)
      XRB=1.
      XPAGE=2.5
      XRB2=6.
      EX=6.
      DO 100 I=1,20
      CALL FA(XRB,1.5,.25)
      CALL PLOT(XPAGE,3.5,2)
      CALL LINK(XPAGE,3.5,EX,6.)
      CALL PLOT(XRB2,1.5,.25)
      XRB=XRB+.5
      XPAGE=XPAGE+.5
      XRB2=XRB2+.5
      EX=EX+.5
100   CALL ERASE
      STOP
      END
```

In this chapter motion with respect to the fixed link will be considered as
absolute motion, diagrammed in Figure 5.4. The display program moves
the four-bar linkage across the screen in .5-unit steps with a total display
of 40 frames. A frame is one generation of a four-bar linkage display; in
order to show motion, the display must be plotted over and over again
along the desired path of travel.

    A display unit is defined inside the subroutine PLOT listed below.
In this listing display units are converted to English inches for ease of
reference; however, the statement IX,IY times 130 could have been set
equal to any SI unit of measurement.

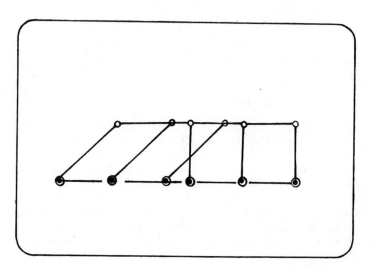

Figure 5.4   Absolute FA motion.

```
C     ****************************************************************
C     *                                                              *
C     *  THIS SUBROUTINE CONVERTS CALCOMP PLOTTER COMMANDS *
C     *  INTO THE PLOT-10 COMMANDS FOR DISPLAY ON THE TEK-  *
C     *  TRONICS 4010 OR 4012 SERIES  TERMINALS.            *
C     *                                                              *
C     ****************************************************************
      SUBROUTINE  PLOT(XPAGE,YPAGE,IPEN)
      IX=X*130.
      IY=Y*130.
      IF(IPEN.EQ.3)GOTO 1001
      IF(IPEN.EQ.2)GOTO 1002
      IF(IPEN.EQ.-3)GOTO 1003
      IF(IPEN.EQ.-2)GOTO 1004
1001  CALL  MOVABS(IX,IY)
      GOTO 44
1002  CALL  DRWABS(IX,IY)
      GOTO 44
1003  CALL  MOVREL(IX,IY)
      GOTO 44
1004  CALL  DRWREL(IX,IY)
  44  RETURN
      END
```

As mentioned, either English units of metric units can be used; however, the CALCOMP plotter uses only English units. The conversion from real-world units of measurement to display units takes place during the PLOT command or any time a display vector, either dark or visible, is generated for the CRT presentation. The subroutine PLOT is used to move the vector generator for all other graphic entities, such as DOT, CIRCLE, CENTER, LABEL, and the like.

During the analysis of a four-bar linkage display, it will become apparent that the absolute motion of any point in a four-bar depends upon the length of the CALL LINK's relative to the length of the CALL CENTER. It is the purpose of this chapter to illustrate that variations of the motions of machines, and different machines, are evolved by

1. Changing the lengths of the links
2. Inversion of the pairs
3. Using sliding pairs
4. Enlargement of pairs

Relative Motion

Since the absolute motion of the fixed link does not change the relative motion of the other links as shown in Figure 5.4, the relative motions can now be displayed and studied. Let is begin with a display program to turn crank AB in Figure 5.5.

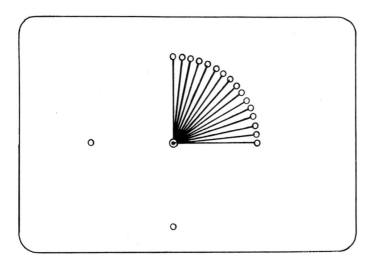

Figure 5.5   Rotation of crank AB.

```
CALL INITT(240)
CALL FA(4.,3.,.25)
CALL ROTATE(4.,3.,1.75,90.,15,6.)
CALL FINITT(0, 0)
STOP
END
SUBROUTINE ROTATE(X,Y,R,SANG,N,THETA)
SANG=(3.14/180.)*SANG
XX=R*(1.-COS(SANG))
YY=R*(SIN(SANG))
DX=X+XX
DY=Y+YY
CALL LINK(X,Y,DX,DY)
THETA=(3.14159/180.)*THETA
THETA1=THETA
D0 200 I=1,N
FEE=SANG+THETA
PX=R*(1.-COS(FEE))
PY=R*(SIN(FEE))
DX=Y+PX
DY=Y+PY
CALL LINK(X,Y,DX,DY)
THETA=THETA1+THETA
200  CALL ERASE
RETURN
END
```

Crank AB moved through 90° of arc in 6° arc segments. The screen was
erased between each link presentation so that motion was simulated for
the machine designer.

Angular Speeds

The laws governing the relative angular speeds of a link were presented
in Chapter 3. A careful study of Figure 5.6 will graphically determine
the law applying to the angular speed of the cranks AB and CD.

  The angular speeds of the cranks of a four-bar linkage vary inversely as the lengths of the perpendiculars or any two parallel
  lines displayed from the fixed centers to the center line of the
  connecting rod; also, inversely as the distance from the fixed
  centers to the point of intersection of the center line of the connecting rod and the line often displayed normal to the centers.

Dead Points

A position in the cycle of motion of the driven crank where a straight
line exists with the connecting rod is known as a dead point. In Figure
5.5 a dead point occurs just before 90° of rotation; therefore, a SANG
(starting angle) of 90° was used for the presentation of Figure 5.6.

Centrodes

In Figure 5.3 A and D are fixed axes, and the elements containing AB,
BC, CD, and AD are labeled 2, 3, 4, and 1 by CALL LABEL. Cranks
2 and 4 oscillate through angles as indicated by Figure 5.6. If the
CALL ERASE statement were removed from the display program, the
linkage would be displayed in a series of positions superimposed one
upon the other. In any display position of ABCD the instantaneous axis
of link 3 is IC(3). It is displayed by plotting the linkage in a series of
positions and displaying a smooth scalar through the successive positions
of IC(3) with CALL SMOOT. This scalar is the locus of the IC(3) for the
range of motion specified in Figure 5.6 and is called the centrode of 3.

FOUR-BAR LINKAGE SITUATIONS

A centrode exists for every position of an element having any coplanar
motion other than pure translation (Figure 5.4) or rotation about a fixed
axis (Figure 5.5). Therefore, every element having coplanar motion has
a centrode. The display of the centrode depends upon the display of the
path of any two points in the element. In the coupler link of a four-bar
linkage it depends upon the relative lengths of the four links. Often, the
display outline of the centrode is a complex curve having little practical
value, but in some special cases it is a simple curve such as a circle,
ellipse, or straight line. If this second-order case occurs, it may be
convenient to display the linkage without CALL ERASE and make use of
the centrode either in analyzing the motion or in constructing the mechanism.

Crank and Rocker

The link AD represents a fixed relationship in Figure 5.6. Suppose that
the crank AB revolves while the lever DC oscillates about its axis D. In
order for this to be displayed as Figure 5.7, the coupler length must be
adjusted to eliminate dead points. The coupler must be increased in length
before the desired motion can be displayed. Figure 5.7 indicates the programmed motion for AB as display points B, B(1), and B(2); lever CD is
C, C(1), and C(2). It can be seen that B(1)AC(1) forms a straight line,
whereas AB(2)C(2) does not. Therefore, if AB is the driver, oscillation
of CD takes place between C(1) at the extreme left and C(2) at the extreme

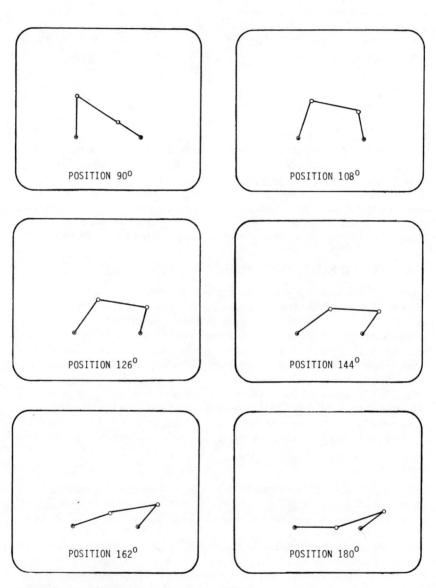

Figure 5.6   Still frames during rotation.

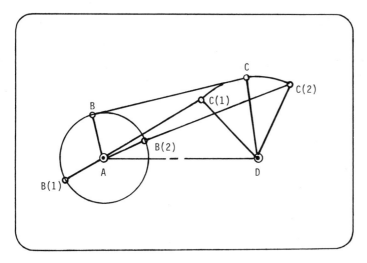

Figure 5.7   Crank and rocker.

right, and no dead points are present. However, if CD is the driver, there
is a dead point at B(1) and the lever cannot turn AB through 360° of rotation.

Drag Link

This example can be programmed so that in Figure 5.7 one of the two longer
links is the fixed link, and the proportions displayed so that only one of the
cranks (AB) makes a revolution motion. In Figure 5.8 the display propor-

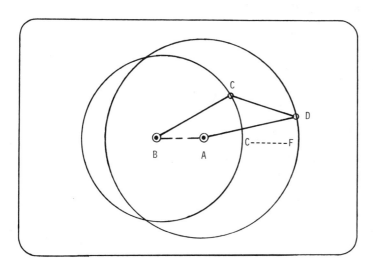

Figure 5.8   Display of drag link.

tions are nearly the same as Figure 5.7, but AB is displayed as the fixed link. The links BC and AD are programmed to revolve about B and A while CD becomes the coupler or connecting rod. This computer-aided design of a machine element is known as the drag link.

In order that the links known as cranks can make complete revolutions on the display creen and that no dead points are displayed, the following program conditions must be input:

1. Each display crank must be longer than the line of centers.
2. Link CD in Figure 5.8 must be greater than segment CF, which represents the distance between the two crank circles.

Parallel and Nonparallel

In Figure 5.9 the crank displayed as AB is equal in length to the crank CD, and the line of centers AD is equal to the connecting rod BC. The following program will display Figure 5.9:

```
       CALL INITT(240)
       CALL BEGIN
C      INSERT THE CALL TO DISPLAY FA1
       CALL FA(2.25,3.5,.25)
C      LABEL THE FIXED ASIX "A"
       CALL LABEL(2.25,3.,.25,'A',0.,1)
C      INSET THE SECOND FIXED AXIS
       CALL FA(8.0,3.5,.35)
C      LABEL THE FIXED AXIS "D"
       CALL LABEL(8.0,3.,.25,'B',0.,1)
C      DISPLAY THE CONNECTING ROD
       CALL LINK(3.5,5.,9.25,5.)
C      LABEL THE CONNECTING ROD "B            C"
       CALL LABEL(3.75,5.25,.25,'B            C',0.,26)
       CALL PLOT(3.5,5.0,2)
       CALL PLOT(9.25,5.0,2)
       CALL FINITT(0,0)
       STOP
       END
```

Using the TS0 instructions from Chapter 2, the program above can be displayed for analysis. The center lines of the linkage are not displayed and can be added to the program by

```
       CALL CENTER(2.3,3.5,8.0,3.5)
       CALL DASH(2.3,3.5,2.3,5.0,.1)
       CALL DASH(1.9,5.0,3.5,5.0,.1)
       CALL DASH(8.0,3.5,8.0,5.0,.1)
```

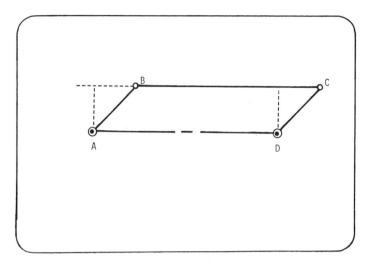

Figure 5.9   CRT display of parallel linkage.

and the display now forms a parallelogram in any position chosen, provided that the input for the animation of the cranks turn in the same sense.
        Careful study of the CRT display indicates that the dashed lines have
formed perpendiculars. These perpendiculars are always equal and the
two cranks are always turning at the same angular speed. If the input to
the cranks provides opposite senses, the perpendiculars do not remain
equal, as shown in Figure 5.10. The display of Figure 5.10 indicates
that while crank AB turns with uniform angular speed, the second crank
CD has a varying angular speed; although both make one complete turn
in the same length of time. The opposite senses of revolution causes
the cranks to have dead points. In subsequent chapters we will discover
how to overcome these dead points by means of special elements placed
at the instantaneous axis of the connecting rod.
        Figure 5.9 represents a parallel and equal case, when the cranks
are turned with the same sense. An opposite sense of the same linkage,
as shown in Figure 5.10, represents a nonparallel, unequal situation.

Equal Crank

The four-bar linkage shown in Figure 5.9 differs from that shown in
Figure 5.7 in that the connecting rod is shorter. This difference has
an important effect on the relative angular motion of the cranks. If the
connecting rod is shorter than the line of center (AD) and is parallel to
it, a principle used in steering mechanisms is displayed.

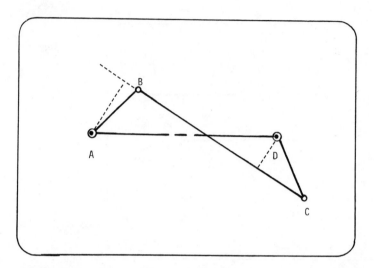

Figure 5.10   Opposite crank revolution.

<u>Automotive Steering</u>.   In Figure 5.11 the linkage is displayed in bright
vectors, ABCD, in the parallel position. The angle formed by BAD is
equal to angle CDA. Now if crank CD is turned through the angle
annotated Ψ, toward the center, then crank AB will turn through a
corresponding angle ψ, which is smaller than Ψ. For a left turn, if

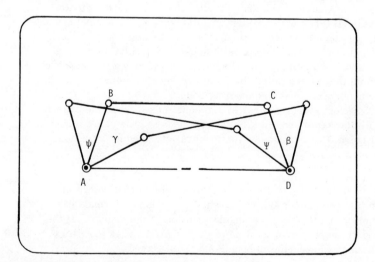

Figure 5.11   Automotive linkage.

crank DC is turned through the angle β away from the center, AB will turn
through the corresponding angle γ, which is larger than β. The steering
of automobiles is programmed and displayed in Figure 5.12. The direction
in which the wireform model moves is controlled by simultaneous turning
of the front wheels about the pivot pins. If the model is making a left turn,
the axis of the left wheel must swing about the pin through a greater angle
than the right wheel. If a right turn is being made, the opposite condition
must exist. The programmed relationship between the swing of the axes
would be such that their center lines, if extended, would always intersect
on the center line of the rear axle, as shown at FA in Figure 5.12. Then
all members of the wireform would be moving about a vertical axis through
FA and the tendency of the wheels to skid would be reduced to a minimum.
      Because of the practical difficulties of using this linkage theory
illustrated in Figure 5.12, most manufacturers employ a linkage of the
type displayed in Figure 5.11. In the assembly and manufacture of a
steering linkage, arms are attached to the short axles shown in Figure
5.12. Suitable mechanisms are designed for connections from the steering column to one of these arms to give the driver control of the linkage.

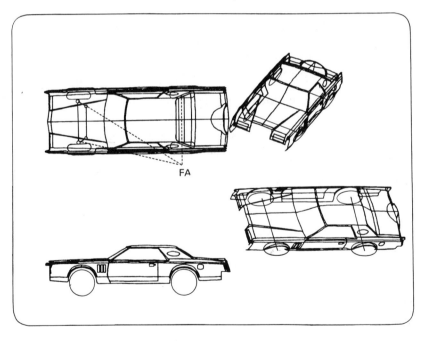

Figure 5.12   Automotive steering linkage. (Courtesy of Automated
Graphics Laboratory, Clemson University.)

Valve Gears.   The four-bar linkage can, if properly proportioned, be
made to produce a slow motion of one of the cranks. Such a combination
is used in Figure 5.13, where two cranks are arranged to turn on fixed
centers and are connected by a link longer than the line of centers. If
one of the cranks is turned clockwise, the opposite crank will turn clockwise but with slower speed. The motion of the second crank becomes
zero when the first crank meets its dead point and causes the opposite
crank to return to its starting position, slowly at first, then gradually
faster. These type of motions are used in the small engine valving shown
in Figure 5.13.

LINKAGE AND SLIDING MEMBERS

In Chapter 1 the relationship between a linkage and a sliding member such
as the one in Figure 1.7 was shown. To this point a linkage was displayed
from the computer database as a rigid element that serves to transmit
force from one element member to another. A linkage display thus far
has also consisted of a number of pairs of display elements connected by

CALL LINK

where the display combination was such that relative motion of the links
was possible, and the motion of each member relative to the others was
definite. In the remaining portion of this chapter one of the links will be
fixed. This then becomes a mechanism.

Referring again to Figure 1.7, the four links of the four-bar
mechanism are AB, BC, CD ∞, and AD ∞ (the lines AD and CD meet
at infinity. Four cases occur in the use of this type of linkage and
sliding member:

1. Connecting rod longer than the finite crank (fixed body)
2. Connecting rod shorter than the finite crank (block fixed)
3. Line of centers longer than the finite crank
4. Line of centers shorter than the finite crank

Slider-Crank

Figure 5.13 represents in CAMD form, the cylinder, piston, connecting
rod, crank, crankshaft, valves, and power cycle of a gasoline engine.
The elements bear the same relationship as those of the corresponding
links in Figure 1.7. The cylinder walls, although not absolutely fixed
to the earth, are fixed to the machine frame and are therefore considered
as the fixed piece of the four-bar linkage. The crankshaft at FA turns in
bearings rigidly attached to the cylinder. The connecting rod is the
coupler of the four-bar, and the piston receives its power from the
exploding fuel and is the driving member of the linkage. In this case
the connecting rod is longer than the crank, placing this machine into
linkage case 1.

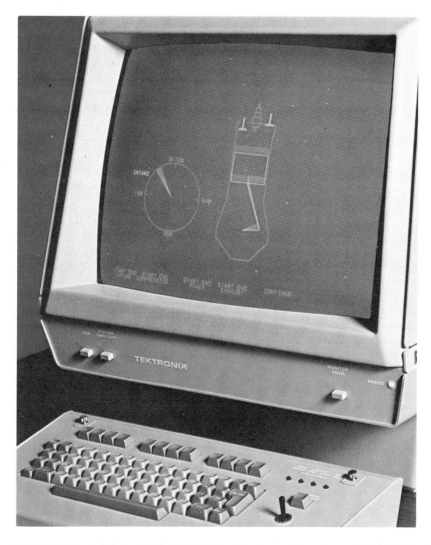

Figure 5.13   Slider-crank mechanism display. (Courtesy of Tektronix, Inc.)

Slider-Slot

Figure 1.7, an example of a slider and a crank, is also a sliding-slot
linkage. If the block inside the slot reaches either end of the slot during
the kinematic cycle, it becomes momentarily a pure slider-crank like
the gasoline engine. The connecting rod may be in constant motion
driven by the crank while the block may be in a period of rest, transmitting its motion to the slider. Many machine parts make use of this

CAMD display technique. The most common are in the areas of
hydraulic pumping applications. A familiar example to many may be
the farm water pump (hand-action type).

The display of slider-slot linkages for CAMD analysis has been
made fairly simple by the use of "write-through" terminals such as
the Tektronix 618. Used in connection with an IBM light pen terminal,
the two hardware devices can place slider-slot mechanisms for quick
display on the DVST (618, direct-view storage tube). Choice of linkages
are listed on the IBM, or a similar menu device, and the light pen
selects the item to be displayed on the DVST. To aid in the placement
or relocation of the linkage, crosshairs are displayed on the face of
the display CRT as shown in Figure 5.14. This tracking display is
part of the write-through capability of the 618 CRT. The designer
positions the crossshairs (cursor by a lever device called a joy-stick.
When a suitable location for the slider-slot linkage has been determined,
the designer selects a preprogrammed display item by pressing the light
pen to the face of the second terminal. The required slider-slot is then
plotted at the location of the cursor on the 618 DVST screen. If in the
process of analysis the designer wishes to relocate the linkage, it may
be moved by use of the joy-stick. Figure 5.13 was displayed in this
fashion. The bright display vectors forming the linkage, valves, and
intake cycle were positioned by the cursor and are part of the write-
through capability of the display terminal. The remainder of the display vectors in Figure 5.13 are stationary (DVST) and were plotted from
the application program written by the machine designer.

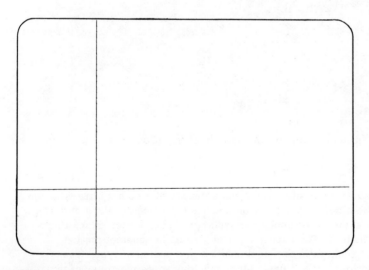

Figure 5.14   Cursor display.

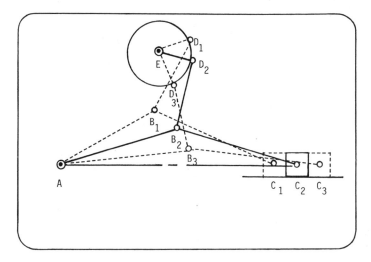

Figure 5.15   Swinging linkage animation.

Swinging Linkage

The swinging linkage shown as a toggle mechanism in Figure 5.15 is
one application for this type of linkage. When the slider C approaches
the end of its stroke, its velocity decreases, causing the slider C to
be capable of exerting a high force when the applied force is low. This
can be demonstrated by taking moments about FA(A) as shown in Figure
5.16. Several variations of the swinging linkage are used in pneumatic
devices, clutches, and presses.

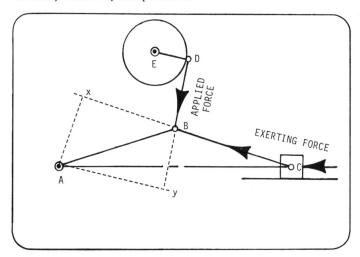

Figure 5.16   Swinging linkage analysis.

Oscillating and Quick Return

Figure 5.13 is a display of the linkage of an oscillating engine. The
combination of oscillating and quick-return mechanisms are common-
place in machine design. Machine tools employ the quick-return,
oscillating-arm mechanism for driving rams. The ram holding the
cutting tool slides back and forth in fixed guides. The oscillating and
quick-return linkage is designed so that the speed of the ram as it
moves over the work piece, removing material, is nearly uniform, and
then the return stroke is made more rapidly.

In the design of new machines containing oscillating and quick-return
mechanisms, designers can write in mathematical models so that the
function of the ram can be calculated, displayed, and analyzed in one
program. A computer program such as this contains a minimum of
three precision points during the travel of the ram. The designer must
also input limits to describe the model.

```
C     ****************************************************************
C     *                                                              *
C     * COMPUTER-AIDED DESIGN OF A 4-BAR FUNCTION GENERATOR*
C     * FOR AN OSCILLATING-QUICK RETURN LINKAGE.             *
C     *                                                              *
C     * X1   = VALUE OF INDEPENDENT VARIABLE AT FIRST        *
C     *          PRECISION PT                                      *
C     * X2   = VALUE OF INDEPENDENT VARIABLE AT SECOND       *
C     *          PRECISION PT                                      *
C     * X3   = VALUE OF INDEPENDENT VARIABLE AT THIRD        *
C     *          PRECISION PT                                      *
C     * Y1   = VALUE OF DEPENDENT VARIABLE CORRESPONDING     *
C     *          TO X1                                             *
C     * Y2   = VALUE OF DEPENDENT VARIABLE CORRESPONDING     *
C     *          TO X2                                             *
C     * Y3   = VALUE OF DEPENDENT VARIABLE CORRESPONDING     *
C     *          TO X3                                             *
C     * PHI  = INPUT ANGLE AT X1 (DEGREES)                   *
C     * PSI  = OUTPUT ANGLE AT X1 (DEGREES)                  *
C     * DX   = RANGE OF INDEPENDENT VARIABLE                 *
C     * CY   = RANGE OF DEPENDENT VARIABLE                   *
C     * DPHI = RANGE OF INPUT ANGLE                          *
C     * DPSI = RANGE OF OUTPUT ANGLE                         *
C     *                                                              *
C     ****************************************************************
```

```
      COMMON Y1,Y3,X1,X3,DPHI,DPSI,DX,DY,PHI,PSI
      DR=3.14159/180.
      READ(1,1)X1,X2,X3,Y1,Y2,Y3,PHI,PSI,DPHI,DPSI,DX,DY
    1 FORMAT(12F5.2)
      CALL INITT(240)
      CALL FSYN(X1,X2,X3,Y1,Y2,Y3,PHI,PSI,DPHI,DPSI,DX,DY,A1,
     +A2,A3,A4)
C     THE PROGRAM NOW BRANCHES TO A SUBROUTINE FSYN WHICH
C     DETERMINES THE LINK PROPORTIONS FOR A 4-BAR FUNCTION
      PRINT 2
    2 FORMAT('1',//,10X,'***** RESULTS *****')
      PRINT 3,A1,A2,A3,A4
    3 FORMAT('0',5X,'CRANK LENGTH ........',F10.3,/,6X,
     'COUPLER LENGTH+......',F10.3,/,6X,'ROCKER
     LENGTH ......',F10.3,/,6X,'FRAME LENGTH +TH ......',
     F10.3)
C     SWITCH SIGNS ON CRANK AND ROCKER TO MAKE NOTATION
C     CONFORM TO LNK OUT SUBROUTINE
      CALL LNKOUT(A4,A1,A2,A3,PHI,PSI
C     THE PROGRAM NOW BRANCHES TO A SUBROUTINE LNKOUT
C     WHICH DISPLAYS A SCHEMATIC DIAGRAM OF THE 4-BAR
C     LINKAGE ON A DVST TERMINAL
      DEL=0.025
      PHIMN=PHI-X1/DX*DPHI
      PHIMX=PHIMN+DPHI
      CALL ERANAL(A4,A1,A2,A3,PHIMN,PHIMX,DEL)
C     THE PROGRAM NOW BRANCHES TO A SUBROUTINE ERANAL
C     WHICH LISTS THE STRUCTURAL ERROR ANALYSIS OF THE
C     4-BAR LINKAGE
      CALL FINITT(0, 0)
      STOP
      END
```

Sample input data to satisfy the READ statement of this program could appear as

6. 45. 84. .0018 .0137 .0256 87. 80. 120. 60. 90. 1.

where $Y1 = SIN(X1*DR)$, $Y2 = SIN(X2*DR)$, and $Y3 = SIN(X3*DR)$. The program then calls a subroutine INITT, which places the graphics terminal in a display mode. Next the program calls a subroutine FSYN to determine the value of A1 (crank length). A2 (coupler length), A3, (rocker length), and A4 (fixed length).

```
C      ***********************************************************
C      *                                                         *
C      *     THIS SUBROUTINE DETERMINES THE LINK PROPORTIONS     *
C      *     FOR THE 4-BAR LINKAGE UNDER CONSIDERATION           *
C      *                                                         *
C      ***********************************************************
       SUBROUTINE FSYN(X1,X2,X3,Y1,Y2,Y3,PHID,PSID,DPHI,DPSI,
      +DX, DY, A1, A2, A3, A4)
       REAL K1,K2,K3,KD
       DR=3.14159/180.
       PHI1=PHID*DR
       PSI1=PSID*DR
       A=DPHI/DX*DR
       B=DPSI/DY*DR
       PHI2=PHI1+(X2-X1)*A
       PHI3=PHI1+(X3-X1)*A
       PSI2=PSI1+(Y2-Y1)*B
       PSI3=PSI1+(Y3-Y1)*B
       W1=COS(PHI1)-COS(PHI2)
       W2=COS(PSI1)-COS(PSI2)
       W3=COS(PHI1-PSI1)-COS(PHI2-PSI2)
       W4=COS(PHI1)-COS(PHI3)
       W5=COS(PSI1)-COS(PSI3)
       W6=COS(PHI1-PSI1)-COS(PHI3-PSI3)
       KD=W2*W4-W1*W5
       K1=(W2*W6-W3*W5)/KD
       K2=(W1*W6-W3*W4)/KD
       K3=COS(PHI1-PSI1)-K1*COS(PHI1)+K2*COS(PSI1)
       A4=1
       A1=1./K2
       A3=1./K1
       A2=SQRT(A1*A1+A3*A3+A4*A4-2.*A1*A3*K3)
       RETURN
       END
```

The FSYN subroutine returns information for the program to print the following results:

```
      ***** RESULTS *****

CRANK LENGTH......        -0.201
COUPLER LENGTH....         1.038
ROCKER LENGTH.....        -0.270
FRAME LENGTH .....         1.000
```

Next, the signs on the crank and rocker are switched to make the notation conform to the LNKOUT subroutine by

        A1=-A1
        A3=-A3

```
C    ****************************************************************
C    *                                                              *
C    * THIS SUBROUTINE DISPLAYS A SCHEMATIC DIAGRAM OF A            *
C    * 4-BAR THE CALLING SEQUENCE CONTAINS THE FOLLOWING:          *
C    *                                                              *
C    *      R1   FIXED LINK LENGTH                                  *
C    *      R2   CRANK LENGTH                                       *
C    *      R3   COUPLER LENGTH                                     *
C    *      R4   ROCKER LENGTH                                      *
C    *      PSI  ANGLE BETWEEN R2 AND HORIZONTAL                    *
C    *      PHI  ANGLE BETWEEN R4 AND HORIZONTAL                    *
C    *           (ANGLES MEASURED POSITIVE IN CCW DIRECTION)        *
C    *           DIMENSIONS ARE AUTOMATICALLY SCALED TO FIT         *
C    *           CRT SCREEN                                         *
C    *                                                              *
C    ****************************************************************
        SUBROUTINE LNKOUT(R1,R2,R3,R4,PSI,PHI)
        DIMENSION R(4)
        RAD=3.14159/180.
        PSIR=PSI*RAD
        PHIR=PHI*RAD
        SAVE=0.0
        R(1)=R1
        R(2)=R2
        R(3)=R3
        R(4)=R4
        DO 1 I=1,4
        IF(SAVE.LT.R(I)) SAVE=R(I)
      1 CONTINUE
        DO 2 I=1,4
      2 R(I)=5.*R(I)/SAVE
        R2X=R(2)*COS(PSIR)
        R2Y=R(2)*SIN(PSIR)
        R4Y=R(4)*SIN(PHIR)
        R4X=R(4)*COS(PHIR)
        YMN=0.
        IF(R2Y.LT.YMN)YMN=R2Y
```

```
        IF(R4Y.LT.YMN)YMN=R4Y
        YH=4.-YMN
        R2Y=R2Y+YH
        R4Y=R4Y+YH
        CALL JOINT(0.0,YH)
        CALL PLOT(0.0,YH,3)
        CALL PLOT(R2X,R2Y,2)
        CALL LABEL(R2X,R2Y,).07,'1',0.,-1)
        R3X=R(1)+R4X
        CALL PLOT(R3X,R4Y,1)
        CALL LABEL(R3X,R4Y,).07,'1',0.,-1)
        CALL PLOT(R(1),YH,1)
        CALL JOINT(R(1),YH)
        CALL LABEL(0.,3.,.14,'FRAME LENGTH......',0.,18)
        CALL NUMBER(2.75,3.,.14,R1,0.,4)
        CALL LABEL(0.,2.5,.14,'CRANK LENGTH......',0.,18)
        CALL NUMBER(2.75,2.5,.14,R2,0.,4)
        CALL LABEL(0.,2.,.14,'COUPLER LENGTH.....',0.,18)
        CALL NUMBER(2.75,2.,.14,R3,0.,4)
        CALL LABEL(0.,1.5,.14,'ROCKER LENGTH......',0.,18)
        CALL NUMBER(2.75,1.5,.14,R4,0.,4)
        RETURN
        END
        SUBROUTINE JOINT(X,Y)
        CALL PLOT(X-.05,Y,3)
        CALL PLOT(X+.05,Y,2)
        CALL PLOT(X,Y-.05,3)
        CALL PLOT(X,Y+.05,2)
        CALL PLOT(X-.11,Y-.19,3)
        CALL PLOT(X-.11,Y-.08,2)
        CALL PLOT(X-.05,Y-.02,1)
        CALL PLOT(X+.05,Y-.02,1)
        CALL PLOT(X+.11,Y=.08,1)
        CALL PLOT(X+.11,Y-.19,1)
        CALL PLOT(X+.17,Y-.19,3)
        CALL PLOT(X-.17,Y-.19,2)
        RETURN
        END
```

The program then sets three input variables for the subroutine
ERANAL:

```
        DEL=0.025
        PHIMN=PHI-X1/DX*DPHI
```

and

```
        PHIMX=PHIMN+DPHI
```

which are used by the subroutine ERANAL to determine the structural
error analysis of the four-bar under consideration. This subroutine
returns

STRUCTURAL FRROR ANALYSIS

INPUT DATA

FRAME LENGTH....	1.000
CRANK LENGTH....	0.201
COUPLER LENGTH..	1.038
ROCKER LENGTH...	0.270
P2MIN	79.000
P2MAX	199.000
DEL	0.026

RESULTS

INPUT	OUTPUT	ERROR
ANGLE	ANGLE	(DEG)
(DEG)	(DEG	

```
C     ****************************************************************
C     *                                                              *
C     * SUBROUTINE FOR THE STRUCTURAL ERROR ANALYSIS OF A            *
C     * 4-BAR                                                        *
C     *                                                              *
C     * ****************************************************************
C     * *                                                        *   *
C     * *     THE USER MUST SUPPLY FUNCTION P TO DETERMINE       *   *
C     * *     THE THEORETICAL VALUE OF THE OUTPUT ANGLE          *   *
C     * *     (RADIANS) FOR INPUT ANGLE VALUE OF P.              *   *
C     * *                                                        *   *
C     * ****************************************************************
C     * *                                                        *   *
C     * THE CALLING SEQUENCE CONTAINS THE FOLLOWING:                 *
C     *                                                              *
C     *     R1              FRAME LENGTH                             *
C     *     R2              CRANK LENGTH                             *
C     *     R3              COUPLER LENGTH                           *
C     *     R4              ROCKER LENGTH                            *
C     *     P2MIN           MINIMUM VALUE FOR INPUT ANGLE            *
C     *                     (DEG)                                    *
C     *     P2MAX           MAXIMUM VALUE FOR INPUT ANGLE            *
C     *                     (DEG)                                    *
C     *     DEL             1./NUMBER OF POINTS TO BE                *
C     *                     EVALUATED                                *
C     *                                                              *
C     ****************************************************************
```

```
      SUBROUTINE  ERANAL(R1,R2,R3,R4,P2MIN,P2MAX,DEL)
      DIMENSION ERR(500),OUT(500),YP(500)
      REAL L,L2
      DIMENSION LABL1(40),LABL2(40)
      DATA LABL1/'    ','    ','    ','    ','CRAN','K AN','GLE ','
     +(DEG','REES',')')    '/
      DATA LABL2/'    ','    ','    ','    ','OUTP','UT E',
     +'RROR','(DE','GREE','S)    '/
      PRINT 8
    8 FORMAT('1',//,20X,'STRUCTURAL ERROR ANALYSIS',//)
      PRINT 9
    9 FORMAT(' ',27X,'INPUT DATA',/)
      PRINT 7,R1,R2,R3,R4,P2MIN,P2MAX,DEL
    7 FORMAT(' ',19X,'FRAME LENGTH....',F10.3,/,
     +20X,'CRANK LENGTH....',F10.3,/,
     +20X,'COUPLER LENGTH..',F10.3,/,
     +20X,'ROCKER LENGTH...',F10.3,/,
     +20X,'P2MIN...........',F10.3,/,
     +20X,'P2MAX...........',F10.3,/,
     +20X,'DEL.............',F10.3,/,
      PRINT 10
   10 FORMAT(' ',29X,'RESULTS',//
      PRINT 11
   11 FORMAT(' ',19X,'INPUT',6X,'OUTPUT',6X,'ERROR',/,20X,
     +'ANGLE',7X,'ANGLE',6X,'(DEG)',/,20X,'(DEG)',//)
      N=1./DEL+1.
      DR=3.14159/180.
      P2MIN=P2MIN*DR
      P2MAX=P2MAX*DR
      DEL=DEL*(P2MAX-P2MIN)
      P2=P2MIN-DEL
      D0 1 I=1,N
      P2=P2+DEL
      L2=R1*R1+R2*R2-2.*R1*R2*C0S(P2)
      L=SQRT(L2)
      B=-ARC0S((R1*R1+L2-R2*R2)/(2.*R1*L))
      T=ARC0S((R4*R4+L2-R3*R3)/(2.*R4*L))
      PI2=2.*3.14159
      IF(P2.LE.PI2. and .P2.GE.3.14159)B=-B
      T4=3.14159-T+B
      TCK=FUN(P2)
      ERR(I)=(T4-TCK)/DR
      OUT(I)=P2/DR
      TT=TCK/DR
      PRINT 12,OUT(I),TT,ERR(I)
```

```
   12 FORMAT(15X,3F11.3)
    1 CONTINUE
      XLNG=5.
      YLNG=4.
      CALL SCALE(OUT,KLNG,N,1)
      CALL SCALE(ERR,YLNG,N,1)
      CALL AXIS(6.0,3.0,LABL1,-40,XLNG,0.,OUT(N+1),OUT(N+2))
      CALL A XIS(6.0,1.0,LABL2,40,YLNG,90.,ERR(N+1),ERR(N+2))
      CALL PLOT(6.,1.,-3)
      CALL FLINE(OUT,ERR,-N,1,0)
      XSET=12.0
      CALL PLOT(XSET,0.,-3)
      CALL FINITT(XSET,0.)
      RETURN
      END
      FUNCTION FUN(PP)
      COMMON Y1,Y3,X1,X3,DPHI,DPSI,DX,DY,PHI,PSI
      DR=3.14159/180.
      P=PP/DR
      DP=((P-PHI)/DPHI*DX*X1)*DR
      F=SIN(DP)
      FUN=(PSI+(F-Y1)*DPSI/DY)*DR
      RETURN
      END
```

Turning Linkages

The fourth type or case of linkage and sliding member occurs when the line of centers is shorted than the finite crank. The general term AB becomes the line of centers in this type of linkage display. Link BC is a crank and so is AD; while CD becomes the coupler. Since BA is shorter than BC, this is a typical case 4 linkage.

Linkages have been grouped into cases and have been presented as:

Case 1   Slider-crank or slider-slot
Case 2   Swinging linkage
Case 3   Oscillating and quick return
Case 4   Turning linkages

SUMMARY

In the computer-aided design of linkages, a link was displayed as a machine part related to a functioning mechanism. A mechanism was defined as a combination of links so connected that the relative motion of any two links compel definite motion in every part of the mechanism.

The four-bar linkage was evolved for display on the face of a CRT.
Short computer programs were presented for:

1. The display of a single link
2. The anotation of links on a CRT
3. Display of fixed axes and scalar values
4. Motion of line of centers (animation)
5. Rotation of a single link
6. Display of linkage systems

During the analysis of a four-bar linkage, it became apparent that
absolute motion of any point in a four-bar depends upon the length of the
CALL LINK's relative to the length of the CALL CENTER. Further
analysis illustrated that variations of the motions of machines, and
different machines, are designed by:

1. Changing the lengths of the links
2. Inversion of the pairs
3. Using sliding pairs
4. Enlargement of pairs

Angular speeds, dead points, and centrodes were presented through
displays of crank and rocker, drag links, and parallel/nonparallel four-
bar linkages. A special application of equal crank presentation was programmed for automotive steering and valve gears. Next, four application
cases for linkages and sliding members were discussed, with special programming for the analysis of oscillating and quick-return linkages. The
application programming presented was intended to serve as a typical
method, not as the only procedure for analyses.

The reader has been introduced to a number of new concepts in
Chapter 5, among them:

1. Conversion software for plotter to CRT programming
2. DVST with write-through capability
3. Dual-screen CRT presentation
4. Dynamic programming for animation of machine parts

BIBLIOGRAPHY

Levinson, I. J., Machine Design. Reston Publishing, Reston, Va., 1978.
Parr, R. E., Principles of Mechanical Design. McGraw-Hill,
    New York, 1970.
Patton, W. J., Kinematics. Reston Publishing, Reston, Va., 1979.
Paul, B., Kinematics and Dynamics of Planar Machinery. Prentice-Hall,
    Englewood Cliffs, N.J., 1979.
Phelan, R. M., Fundamentals of Mechanical Design. McGraw-Hill,
    New York, 1970.

Ramous, A. J., Applied Kinematics. Prentice-Hall, Englewood Cliffs,
    N.J., 1972.
Ryan, D. L., Computer-Aided Graphics and Design. Marcel Dekker,
    New York, 1979.
Spotts, M. F., Design of Machine Elements. Prentice-Hall, Englewood
    Cliffs, N.J., 1978.
Wilson, C. E., and W. J. Michels, Mechanism ... Design-Oriented
    Kinematics. American Technical Society, Chicago, 1969.

6

Computer Generation of Transmission Paths

The computer generation of transmission paths can be displayed when the driving member of a computer-aided mechanism is modeled in direct contact with the driven element. The display constituting the driver and follower are programmed either in pure rolling contact or in sliding contact. A display exception to this may occur when the program finds that the lines of contact on both the driver and follower are moving along a display line common to both contact surfaces.

TRANSMISSION OF MOTION BY COMPUTER MODELING

In this chapter several programming segments and basic modeling constructions concerning the two display methods are presented. In the remaining chapters, the application of these display methods is presented for the computer-aided design of cams, gears, friction drives, and types of connectors. The term computer modeling means that machine elements are described by shape, size, and turning axis in computer memory. Two or more elements can then be recalled to display a mechanism containing a driver and follower. The driver is turned about its axis and displayed, together with the location of the follower throughout its path of motion. Almost unlimited types of drivers and followers can be described, displayed, and analyzed using this technique. A large host computer such as the IBM 370 Model 3033 is required, together with the proper display screens, such as the Tektronix 618.

Sliding Contact and Speed Ratios

In the display program for Figure 6.1, one of the fundamental cases of a driver and follower can be studied. FA_1 and FA_2 are fixed axes for the driver and follower and are displayed as

FA1=350,100

and

FA2=350,250

```
//DRYAN JOB (0923-1-410-DR-7,:06,2),'RETURN TO BOX 38'
//STEP1 EXEC TEK,PDS='DRYAN.DISPLY',NAME=DRIVER
//C.SYSIN DD *
      PRINT 100
100   FORMAT('WHAT SPEED RATIO IS DESIRED?SELECT
      +BETWEEN 10 AND 240.')
      READ(1,101)NBAUD
101   FORMAT(I3)
      CALL INITT(NBAUD)
      CALL DRIVER(X1,Y1,RADIUS)
      CALL FOLLOW(X2,Y2,R1,R2,CD)
      CALL FINITT(0,0)
      STOP
      END
/*
//
```

where $X1 = 350$ tekpoints on the display screen and $Y1 = 100$. This positions FA_1 while $X2$ and $Y2$ locate FA_2 of the follower. The subroutines that describes the shape, size, and turning axis are

```
      SUBROUTINE DRIVER(X1,Y1,RADIUS)
      CALL CIRCLE(X1,Y1,RADIUS,270.,30,6.)
      CALL CIRCLE(X1,Y1-RADIUS,RADIUS*2,90.,15,6.)
      CALL PLOT(X1+RADIUS*2,Y1-RADIUS*2,2)
      CALL CIRCLE(X1-RADIUS/2.,Y1-RADIUS*2.,RADIUS/2.,180.,30,6.)
      CALL CIRCLE(X1,Y1-RADIUS*2.,RADIUS,90.,15,6.)
      RETURN
      END
```

and

```
      SUBROUTINE FOLLOW(X2,Y2,R1,R2,CD)
      CALL PLOT(X2-R1,Y2,3)
      CALL PLOT(X2-R2,Y2+CD,2)
      CALL CIRCLE(X2,Y2+CD,R2,0.,30,6.)
      CALL PLOT(X2+R1,Y2,2)
      CALL CIRCLE(X2,Y2,R1,180.,30,6.)
      RETURN
      END
```

In Figure 6.1, machine elements 1 and 2 are turning about fixed axes FA_1 and FA_2 by use of the subroutine ROTATE used for the rotation of crank AB in Figure 5.5. The CALL DRIVER command is substituted for the CALL LINK command and the machine element is then rotated on the display screen. Similarly, the follower can be rotated by the substitution of CALL FOLLOW. The display point labeled P on machine element 1 is, at the instant of display rotation, in contact with an opposite point, labeled K on machine element 2, while element 1 is rotated clockwise. This will cause the follower to turn counterclockwise.

The display rate NBAUD may be varied from 10 to 240. This represents the angular speed of 1 and 2. Then the velocity of P is equal to NBAUD*(FA_1*P) and can be represented by adding a display vector, as shown in Chapter 3. This vector is perpendicular to FA_1P, while the direction of the velocity of K is perpendicular to FA_2K. The magnitude of the velocity of sliding P on K is displayed by changing the baud rate argument of CALL INITT(NBAUD). The profiles of the driver and follower can be modified by changing the computer models stored under the subroutines DRIVER and FOLLOW.

It is apparent that the computer-generated mechanism display of Figure 6.1 can transmit motion only through a limited range of angular motion, and that the driver and follower turn in opposite senses. In

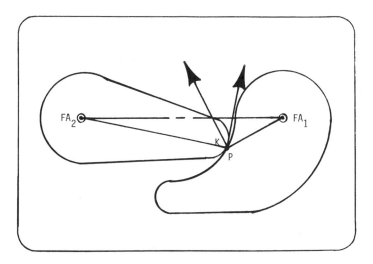

Figure 6.1   Display of driver and follower.

Figure 6.2 a new program was written so that the driver may make a
complete revolution. This will then cause the follower to rotate in the
same direction or oscillate through an angle depending upon where the
subroutines are placed (line of centers, distance, and sizes).

The crank-and-slot drive of Figure 6.2 produces nonuniform
circular motion in which the velocity variation can be manipulated by
a computer subroutine. A display program can be written so that the
driven slot link will present a complete revolution for each revolution
of the input crank shown in Figure 6.2. The choice of output velocity is
dependent on center distance (CD), which is expressed in FORTRAN as
the crank radius (CR) multiplied by a factor (RZ):

    CD=CR*RZ

where RZ is less than 1. If ZR is greater than 1, the slot will oscillate
instead of rotate. When CD = CR*.25, the maximum velocity will be seven
times the minimum. When CD is adjustable as an argument in a display
subroutine, it is possible to change the magnitude of nonuniformity to meet
the changing requirements.

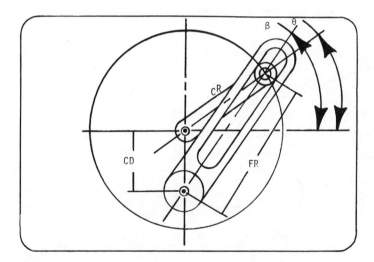

Figure 6.2   Computer-generated mechanism.

```
C    ********************************************************************
C    *                                                                  *
C    * DISPLAY PROGRAM FOR ONE VELOCITY CYCLE PER REVOLU-  *
C    * TION BASED ON THE SLOTTED CRANK DRIVER WHEN THE        *
C    * FOLLOWING ARE KNOWN:                                           *
C    *                                                                  *
C    *      CD = DISTANCE BETWEEN CENTERS                       *
C    *      CR = RADIUS OF DRIVER                                   *
C    *      FR = RADIUS OF FOLLOWER                              *
C    *   THETA = DISPLAY ANGLE OF DRIVER                         *
C    *    BETA = DISPLAY A NGLE OF FOLLOWER                    *
C    *     DIA = THICKNESS OF FOLLOWER                           *
C    *   WIDTH = THICKNESS OF DRIVER                             *
C    *                                                                  *
C    ********************************************************************
         CALL INITT(240)
C    CALL SUBROUTINE TO DISPLAY OUTPUT LINK AT SELECTED
C    SCREEN LOCATION
         CALL OUTSHF(XSCRN,YSCRN,DIA,FR,BETA)
C    CALL SUBROUTINE TO DISPLAY INPUT LINK AT SELECTED
C    SCREEN LOCATION
         CALL INSHFT(XSCREN,YSCREN,WIDTH,CR,THETA)
C    CALL NOTATION AT PROPER LOCATIONS
         CALL LABEL(XSCRN-1., YSCRN+.5,  .2,'CD',0.,2)
         CALL LABEL(XSCRN+2., YSCRN+2., .2,'β',  0.,1)
         CALL LABEL(XSCRN+2.5,YSCRN+2., .2, 'θ',0.,1)
         CALL LABEL(XSCRN+1., YSCRN+.5, .2,'FR',0.,2)
         CALL LABEL(XSCRN+.5, YSCRN+1., .2,'CR',0.,2)
C    CALL DISPLAY GEOMETRY TO INDICATE CENTER LINES,
C    DIMENSION LINES, AND ROTATION OF MACHINE ELEMENTS.
         .
         .
         .
         .
         .
         STOP
         END
```

Of course, the arguments in the program will need to be satisfied with
desired sizes and screen locations for the program to display a static
image as shown in Figure 6.2. When displaying sliding contacts and

speed ratios, the machine designer first writes a short description of
the two elements in contact. A description of the follower in Figure
6.2 would appear as

```
SUBROUTINE OUTSHF(XSCRN,YSCRN,DIA,FR,BETA)
CALL FA(XSCRN,YSCRN)
CALL CIRCLE(XSCRN,YSCRN,DIA/2.,0.,60,6)
X=FR*(XCSRN+(1-COS(BETA))
Y=FR*(YCSRN+(SIN(BETA))
CALL CIRCLE(X,Y,DIA/2.,BETA,60,3.)
CALL PLOT(XSCRN+DIA/°.,YSCRN,2)
CALL PLOT(XSCRN-DIA/2.,YSCRN,3)
CALL PLOT(X,Y,2)
XR=XSCRN+(1-COS(BETA)*FR)
YR=YSCRN+(SIN(BETA)*FR)
CALL CIRCLE(XR,YR,DIA/4.,90.-BETA+180.,60,3.)
CALL PLOT(XSCRN+(1-COS(BETA))*FR,YSCRN+(SIN(BETA))*FR,2)
CALL CIRCLE(XSCRN+(1-COS(BETA))*FR,YSCRN+(SIN(BETA))*FR,
+BETA+180.,60,3.)
CALL PLOT(XR+DIA/4.,YR,2)
RETURN
END
```

where XSCRN and YSCRN are the locations, in display units, for the fixed
axis of the follower. The first CALL statement plots a symbol for an axis
at this location. This display subroutine is considered to be static in that
a single machine member is plotted each time it is called. A program
segment to display several links might be

```
CALL OUTSHF(2.5,1.5,1.,3.,45.)
CALL OUTSHF(1.5,5.5,.875,3.5,0.)
CALL OUTSHF(6.5,2.0,1.25,3.,90.)
```

displayed as shown in Figure 6.3. This illustrates the flexibility the
machine designer may use in the placement of elements. The designer
controls the distance between centers (CD) by placement of CALL
OUTSHF and CALL INSHFT, shown in Figure 6.2. Static displays are
not as useful as dynamic displays; therefore, the designer might use this
segment of a program to show the turning of the slotted link.

```
BETA= 0.
D0 99 I=1,60
CALL OUTSHF(3.5,2.75,1.,2.,BETA)
CALL ERASE
BETA=BETA+6.
99 CONTINUE
```

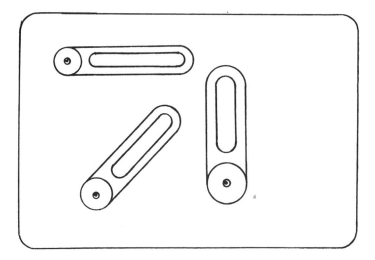

Figure 6.3   Static placement of OUTSHF.

This program segment is designed to display a dynamic image on a direct-view storage tube (DVST) terminal through one revolution. If more revolutions are desired—for example, 10 cranks are to be displayed—the D0 LOOP can be changed to

```
    D0 99 I=1,600

99 CONTINUE
```

where BETA is changed from 0° outside the D0 LOOP to 600 times 6°, or a total of 3600°, resulting in 10 revolutions of the crank. The same program logic can be used to turn the pair of display elements by inserting the call for the second driving link:

```
    THETA=42.
    BETA=54.
    D0 99 I=1,60
    CALL INSHFT(5.,3.5,.750,3.,THETA)
    CALL OUTSHF(5.,1.875,1.25,4.5,BETA)
    CALL ERASE
    THETA=THETA+6.
    BETA=BETA+6.

99 CONTINUE
```

This program segment will display and turn the elements shown in
Figure 6.2.*

Further modification of the program can be made to display the
kinematic relationships of displacement, velocity, and acceleration.
The angular displacement of the follower in Figure 6.2 in relationship
to the driver is determined by

TAN(THETA-90.)=SIN(THETA-90.)/RZ+COS(THETA-90.)

and the displacement is routed to a plotter for selected values of RZ as
shown in Figure 6.4. The angular velocity of the follower is found by
differentiation:

DELTAT/DELTAB=OMEGA*(1+RZ*COS(THETA-90.))/1+RZ**
2+2*RZ*COS(THETA-90.)

These relationships are not FORTRAN expressions but are still in
equation form. Both the angular displacement and the angular velocity
must be set in FORTRAN statements to eliminate computations on the
left-hand side of the equals signs. The angular velocity (OMEGA) of
the follower is plotted as a graph in Figure 6.5.

The acceleration of the follower is found by differentiating the
second relationship for velocity. The acceleration for the values of
RZ = 0.25, 0.50, and 0.75 are plotted in Figure 6.6.

Pitch Points, Angles of Motion, and Pressure Angles

Display points P and K of Figure 6.1, where the resultant of vectors
FA_1P and FA_2K are located, together are called the pitch point. The
line formed by the resultant is called the normal line. The total angle
that the driver turns during the display in contact with the follower is
the angle of action of the driver. The follower also has an angle of
action. With a constant baud rate to simulate angular speed, the magnitudes of the angles of action for both the driver and follower are
direct ratios of their display speed to the desired THETA and BETA
selections together with CD. In the case of variable angular speed
ratios, such as in Figure 6.2, the angles of action are directly proportional to the average baud rate selected and the RZ factor.

In Figure 6.1 the driver and follower were displayed from set
subroutines. These subroutines did not have variable points of rotation as do the Figure 6.2 THETA and BETA arguments. A program
segment was used to turn Figure 6.2, a rather simple method. In the
case of Figure 6.1, another subroutine, ROTATE, turned the static
display. This technique, called subroutine nesting, is not as direct

*Review Chapter 3 for a complete discussion of combining both refresh
and storage graphics on the same display screen for static and dynamic
modes to achieve the velocity and acceleration analysis at reduced cost.

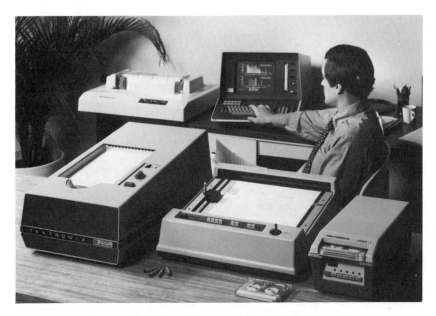

Figure 6.4   Displacement routed to a plotter. (Courtesy of Tektronix, Inc.)

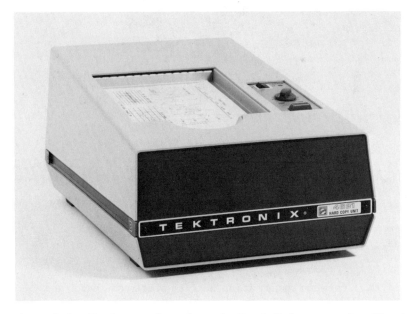

Figure 6.5   Hard copy of angular velocity plotted as a graph. (Courtesy of Tektronix, Inc.)

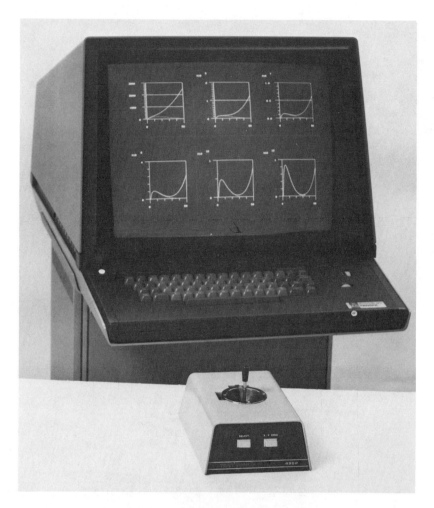

Figure 6.6   Display of the acceleration of the follower. (Courtesy of
Tektronix, Inc.)

and requires additional computational time to present. Both methods
are used. In Figure 6.1 a rather odd-shaped driver was needed, so the
designer wrote a simple subroutine to describe its static shape and then
used the "nested" software approach to make it dynamic. The follower of
Figure 6.1 is a common shape, and in the case where both driver and
follower are symmetrical; the variable approach is best. If in Figure 6.1
both elements were the shape of the follower, the angular speeds of follower
and driver are inversely proportional to the distances from the fixed axes

to the display point P. Display P is then where the normal line intersects the line of centers between FA_1 and FA_2. In other words, if the display were stopped anywhere during the turning period where contact existed; a common resultant could be displayed for FA_1P and FA_2K. This means that the shapes of these contacting surfaces are such that their normal line cuts the line of centers through point P, and the angular speed ratio of the follower and driver will remain constant. In these cases the angle the normal line makes with a perpendicular to the line of centers through the pitch point is called the PRESSURE ANGLE.

Once the pressure angle is known, the path representing the period of contact can be displayed separately. This is important because when the contacting surfaces of the driver and follower are so shaped that their common normal lines intersect the line of centers at the same point for all positions of the elements throughout their angles of action, this results in a constant angular velocity. The curves which are separated by the display program to show these mating surfaces are said to be conjugate curves. It is along these conjugate curves that much useful work is done by machine elements. In Chapter 7 it will become apparent how gears and cams make use of this concept.

PLOTTING THE CONJUGATE TO A GIVEN PATH

Whenever the driver and follower relationship for two machine elements is displayed from computer subroutines, the outline of those parts moved by sliding contact is known. In the case of gears and other elements it is required to display the conjugate or working curve. This curve is determined from the angular speeds of the driver and the follower. The center distance (CD) is also known because the design assigns this relationship by giving values to XSCREN and YSCREN for the driver and follower. As indicated earlier in the chapter, if the angular speed ratio remains constant, the common resultants must cut the line of centers at a point (P). The first step in displaying a working curve, then, is to locate and display P. This is determined by the known value of the angular speed of the driver divided by the angular speed of the follower from the equation

$$\text{OMEGA}_1/\text{OMEGA}_2 = FA_2{}^*P/FA_1{}^*P$$

This is clearly shown in Figure 6.1. Figure 6.1 may be displayed in a static mode on a DVST with write-through capability. For now, a refresh mode must be used to select points on the sliding surfaces and construct normals. If the sliding surfaces produce a curve whose properties are known, such as an arc of a circle, ellipse, or involute of a circle, the normals may be displayed on a DVST with the write-through mode; otherwise, the sense of the normals must be estimated by trial and error with a light pen and refresh terminal, as shown in Figure 6.7.

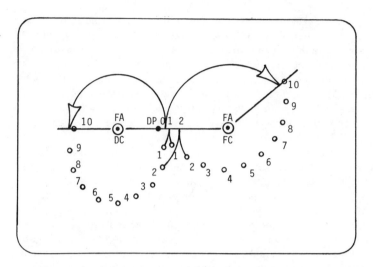

Figure 6.7   Display of a curve in rolling contact.

Rotation with Angular Speeds

For a driver and follower that have a constant angular speed ratio through direct contact, as in the modification of Figure 6.1, the range of motion is limited to a relatively small part of a crank revolution. The working curve displayed is also short, for both Figures 6.1 and 6.2. This condition is essentially the same for the constant and variable angular speed mechanisms discussed earlier in the chapter. Although it might be possible to display the elements of a mechanism with working curves that last through a complete revolution of at least one of the elements, the resulting mechanism actions are of questionable value. When continuous action of some sort is required, a series of duplicate elements are designed on the same axis. For example, if in Figure 6.1 several calls were made to display the driver several times around FA_1 and a corresponding series of calls to duplicate followers around FA_2, then when one pair of elements had ceased to act, the next pair would come into contact. The number of elements on the two axes FA_1 and FA_2 must be determined and spaced close enough to allow one pair to come into action just following the preceding pair. This is the basic principle underlying the design of gears, racks, and gear teeth presented in Chapter 7.

Rolling Contact

If one element of a pair is in contact with another and the motion is such that sliding does not occur between display points along the line of contact, the pair of elements is said to be in rolling contact. The passing surfaces may be of various forms, provided that the basic relationship causes no slipping. For this to happen, every point on one element that is on the line of contact must have the same velocity as the matching point on the other element of the pair. In Figure 6.1, where the driver is in contact with the follower on one side of the line of centers, K is sliding relative to the matching point P. In Figure 6.2, where the contact is along the slot of the follower, sliding also exists, owing to the fact that the sense of the relative velocity of the matching points has changed.

In the modification of Figure 6.1 so that symmetrical driver and follower are displayed, FA_1 and FA_2 have the same velocity and for the angle of action, sliding is zero. Therefore, in that period the driver and follower are in rolling contact. If Figure 6.2 were modified, the same condition could exist, although the shapes of the links happen to be such that the difference between the lengths of the two working curves is not very great. In general, the contact surfaces of driver and follower can be shaped so that they will be in contact on the line of centers at all times and rolling contact will result if no slipping occurs. In such cases, the lengths of the working curves would be the same. Therefore, the conditions for rolling contact between pairs turning on parallel axes which are fixed relative to each other are as follows:

1. The point of contact must always be on the line of centers.
2. The lengths of the contacting surfaces, as displayed on a plane perpendicular to their axes, must be equal.

If the point of contact is at the same place on the line of centers, the angular speed ratio remains constant. When parallel axes are displayed, only circular cylinders will meet the display requirements for rolling contact without slipping plus constant angular speed ratio. Then, friction must be relied on for transmission of motion from driver to follower. In other cases, a circular cylinder and a plane surface might be displayed or, if nonparallel axes are used, a right circular cone along a plane surface. For display of rolling contact, two right circular cones can be displayed for axes intersecting at 90°. For variable speed ratios, an unlimited number of subroutine forms may be displayed, although few will be capable of dynamic revolution of both driver and follower:

DISPLAY OF A CURVE IN ROLLING CONTACT

A computer display of a curve in rolling contact consists of a cathode ray tube (CRT) image or other such output where the relative motion of two lines or surfaces can be shown. In Figure 6.7 the driver turning about the fixed axis FA_1 is displayed in the sense indicated by the arrow. To find the outline of the follower turning about fixed axis FA_2 which will roll without slipping along the display curve DC_0 through DC_{10}, a program is written. This display program uses the two principles stated

earlier: the point of contact must be on the line of centers, and the
lengths of the two curves that come in contact at a given time must be
equal. The display program divides the display curve DC_0-DC_{10} into
parts so small that the length of the arc is approximately equal to the
length of its chord. DC_0 is one display point common to both curves.
With FA_1 as a center radius, an arc is displayed through the first point
of division, DC_1. This arc cuts the line of centers at DP_1 (first display
point). Through this display point (DP_1) an arc is displayed to locate the
first point on the second, or follower, curve FC_1. This is done by a dark
vector calculation to determine the radius distance from DP_1 to FA_2.
This radius is matched with radius DC_0DC_1 to locate the required point
FC_1 on the follower curve.

    Next, the display program repeats the process through a D0 LOOP
to display a dark vector about FA_1 with radius DC_2 cutting the line of
centers at DP_2. Through DP_2 an arc is displayed as DC_2FC_2. From
FC_1 an arc is calculated with radius equal to the chord DC_1DC_2, which
cuts DP_2FC_2 and locates point FC_2 on the follower curve. This process
is CONTINUED by the D0 LOOP for each of points DC_3 through DC_{10},
which in turn display FC_3 through FC_{10}. Now the program can display
a smooth curve through the points by

    CALL SMOOT(XPAGE,YPAGE,IPEN)

where XPAGE and YPAGE are the coordinate locations of the display
point. SMOOT is a graphics subroutine that automates the use of a
french curve. This subroutine displays a smooth curve through a set
of data points. It accomplishes this by using a modified spline-fitting
technique. SMOOT receives a single XPAGE and YPAGE coordinate
pair on each call and will accumulate the curve points until it has received a sufficient number to compute a pair of cubic parametric equations for a smooth curve. This accumulation method requires the
program to specify an initial and a terminal call to the subroutine.
These are placed in IPEN to initialize SMOOT, IPEN = 0, and to
terminate SMOOT, IPEN = -24.

Relative-Motion Display Subprograms

In all the figures illustrated thus far, the normal common to the contacting surface cuts the line of centers between the fixed axis and displays
machine elements that rotate in the opposite sense. Condition 1 for
rolling contact states that the point of contact must always be on the
line of centers. It does not state that the cutting must happen between
the fixed axes. Therefore, if the machine elements were designed such
that the common normal cuts the line of centers on the same side of both
axes, the elements will turn in the same sense.

    Because this condition exists, together with others, subprograms
are developed so that branching may occur in the display program. An

interactive display program guides the machine designer during program execution by prompts. These prompts appear at the top of the
work screen so that the machine designer may select the type of driver
and follower sense.

```
C    ****************************************************************
C    *                                                              *
C    * INTERACTIVE DISPLAY PROGRAM FOR ROLLING CONTACT             *
C    * BETWEEN TWO MACHINE ELEMENTS INVOLVING COPLANAR            *
C    * MOTION ABOUT PARALLEL AXES RELATIVELY FIXED AT A           *
C    * FINITE DISTANCE FROM EACH OTHER.                            *
C    *                                                              *
C    ****************************************************************
C    PROGRAM HEADING
     PRINT, 'INTERACTIVE DISPLAY FOR PURE ROLLING CONTACT'
C
C    PRINT, 'PLEASE ENTER DESIRED BAUD RATE'
     READ(1,100) JBAUD
 100 FORMAT(I3)
     CALL INITT(JBAUD)
C
  99 PRINT, 'PLEASE ENTER X AND Y LOCATION FOR FA(1)'
C
     READ(1,101)XPAGE,YPAGE
 101 FORMAT(2F10.3)
     CALL FA(XPAGE,YPAGE)
C
     PRINT, 'IS THIS LOCATION CORRECT?INPUT YES OR NO.'
C
     READ(1,102)IANSR
     IF(IANSR.EQ.NO)CALL ERASE,G0T0 99
C
  98 PRINT, 'PLEASE ENTER X AND Y LOCATION FOR FA(2)'
C
     READ(1,101)XPAG,YPAG
     CALL FA(XPAG,YPAG)
C
     PRINT, IS THIS LOCATION CORRECT? INPUT YES OR NO.'
C
     READ(1,102)IANSWR
 102 FORMAT(A3)
     IF(IANSWR.EQ.NO)CALL ERASE, G0T0 98
C
     PRINT, 'HOW MANY DISPLAY POINTS ARE LOCATED ABOUT FA(1)?
     INPUT 1 TO 20'
```

```
C
      READ(1,103)NDC
  103 FORMAT(I2)
C
      PRINT, 'INPUT X AND Y LOCATION OF DC(0).'
C
      READ(1,101) XPAG0,YPAG0
      CALL SMOOT(XPAGO,YPAGO,0)
      D0 200 I=1,NDC
      READ(1,101) XPAG0,YPAG0,2)
  200 CONTINUE
C
      PRINT, 'ENTER X AND Y LOCATION OF LAST DISPLAY POINT.'
C
      READ(1,101) XPAE,YPAE
      CALL SMOOT(XPAE,YPAE,-24)
```

Other Subroutines

The program segment just presented was confined to the simplest case, involving only coplanar motion about parallel axes, relatively fixed at a finite distance from each other; it does not show a complete working program. The interactive techniques shown are, however, typical in their nature and with proper modifications will apply equally well if either machine element has rectilinear motion. In this case the distance between the axes is infinite.

In Figure 6.7 the angle θ will be the angle turned by FA_2; FA_1 turns through the angle ρ. Action between FA_1 and FA_2 machine elements ceases when DC_{10} and FC_{10} meet on the line of centers. If the outline of machine 1 were assumed for the remainder of its cycle (360° motion) and the corresponding curve found for the follower, there would be no assurance that the follower would complete its cycle (360° motion) in the same time as the driver. Therefore, if the rolling motion is to be continuous, the given outline (display points for FA_1) may not be chosen at random. The display program does not check for random points entered by a novice CAMD user. To overcome long computation and graphic presentations of unwanted DC data, frequent prompts are printed asking the designer if the locations are correct or if the profile is desirable. By inserting the NO command, the user may erase the screen and select new profile data. This feature is useful when working on a screen with selective erase, but when a DVST terminal is used, the entire display is lost by the erase command. To overcome this limitation the display program must cycle back through all its previous FORTRAN statements and redisplay the screen area. This is very costly in CPU time. Although DVST terminals are very inexpensive, the recycling time required in a

large interactive program for CAMD purposes used on a daily basis
makes the DVST terminal more costly than the selective erase terminal.

Other subroutines are also needed for another series of mechanisms
involving transmission of motion by direct rolling contact. These cases
arise when the two machine elements do not have coplanar motion. Some
of these cases are presented in the next section in connection with rolling
cones, hyperboloids, and cylinder cams.

MACHINE ELEMENTS IN ROLLING CONTACT

The display of rolling contact between machine elements consists of an
application program whereby the relative motion of two lines or surfaces
is presented as a series of consecutive points. The points or elements of
one come successively into contact with those of the other in their order of
display. As illustrated earlier in the chapter, there is no slipping between
the two surfaces, which are programmed to have pure rolling contact. This
is done by displaying all points in contact with the same linear speed.

Two or more bodies representing machine parts may be rotating on
separate axes, yet so arranged that by pure rolling contact, one may cause
the other to turn with an angular speed that is displayed as a ratio of the
angular speed of the first or driver. This speed ratio may be constant or
variable, depending on the subroutines that displayed the two elements. If
the display subroutine CYLIND* is used to prevent the rolling members,
a constant speed ratio will result. Further, if the two members displayed
from CYLIND are equal in size, the transfer of speed from driver to
follower is 1:1. If the driver is larger than the follower, the speed ratio
causes the follower to turn at a greater speed than the driver. Small
drivers with large followers cause slower moving speeds of turning for
the follower.

The axes of the machine elements in rolling contact may be parallel
in the case of CYLIND, with the use of subroutine CONE, or they can be
intersecting. Constant speed ratios similar to CYLIND also result from
these types of displays. Parallel axes connected by elements of irregular
outlines such as those displayed in Figure 6.1 will produce variable speed
ratios. Variable speed ratios can also be produced by nonparallel, nonintersecting axes connected by subroutines that display hyperboloids.

Cylinders and Cones

Figure 6.8 illustrates opposite (case A)- and same-sense displays for
cylinders. In case A, assume that the shafts are held by the frame so
that their centers are at a distance apart just equal to the sum of the
radii of the two cylinders. The display point representing the surfaces

*Display subroutines CYLIND and CONE are described in Ryan (1979),
Chap. 3).

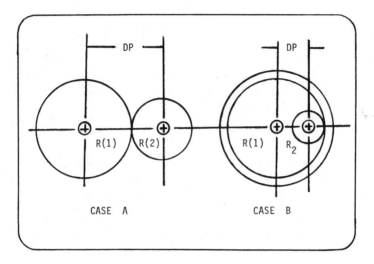

Figure 6.8   Sense displays for cylinders.

that touch at DP can be calculated by the display program from the
statement

    DP=R(1)+R(2)

where DC is the point on both surfaces of the cylinders where no slipping
can occur.

    It was shown earlier that the surface speed of the driver (DC) must be
equal to the surface speed of the follower (FC). In addition, DC and FC
must turn in directions relative to each other so that the elements on DC
(DC_0 through DC_{10}) are in contact with FC elements (FC_0 through FC_{10}).
This causes the senses of rotation for the cylinders in case A of Figure 6.8
to be opposite. If cylinder DC makes N revolutions in a given period of
time and cylinder FC makes M revolutions in the same period, the
surface speed of DC and FC can be found in the display program by the
statements

    SSDC=2*3.14159*R(1)*N

and

    SSFC=2*3.14159*R(2)*M

If DP, N, and M are known, the diameters of the two cylinders may be
found by execution of these statements. The display program can be used
to show that the angular speeds of two cylinders that roll together without
slipping are inversely proportional to the radii of the cylinders.

In case B of Figure 6.8, cylinder DC is dished out with FC turning inside it, so that the contact is between the inner surface of DC and the outer surface of DC. This is displayed as internal rolling of cylinders. The two cylinders will turn in the same sense. In case B, the distance between centers is calculated by the display program by

DP=R(1)-R(2)

instead of R(1)+R(2).

In the display of Figure 6.8, the shafts were necessarily parallel. It is often required to connect two shafts that lie in the same plane but make an angle with each other. This is done by means of right cones or sections of cones, as illustrated in Figure 6.9. The same program logic is used for the ratio of speeds at the base of the cones as for the circles representing the cylinders in Figure 6.8. The cones have a common apex labeled OP and the ratio is expressed as

N/M = R(2)/R(1)

where R(2) = OP*SIN(THETA) and R(1) = OP*SIN(BETA) and

$$\frac{R(2)}{R(1)} = \frac{OP \sin \theta}{OP \sin \beta} = \frac{SIN(THETA)}{SIN(BETA)} = \frac{N}{M}$$

Therefore, the angular speeds of two cones rolling together without slipping are inversely proportional to the sines of the angles. The angle THETA is the cone or center algel of cone DC, and BETA is

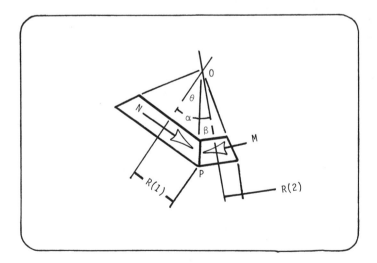

Figure 6.9   Sense display for cones.

the cone angle of FC. The angle ALPHA is the angle between shafts or
shaft angle.

If it is assumed that the sense of rotation is the sense in which a
given shaft is seen to be turning as the user or designer displays the
shaft toward the intersection of the axes, then there are two cases:

1. Opposite sense or rotation of shafts
2. Same sense of rotation of shafts

Two cones may roll in external contact or internal contact. External contact does not mean the opposite sense of rotation and internal
the same sense as in Figure 6.8. As shown in subsequent chapters, the
type of contact (external or internal) depends upon the particular combination of angle between shafts, sense relation, and speed ratio.

Friction Gearing

When used to transmit force, rolling cylinders and cones can be displayed
as friction gearing. The axes are displayed so that they can be specified
to be pressed together with considerable force during manufacture, to
prevent slipping. The surfaces of contact should be designed with slightly
yielding materials, such as plastic, nylon, rubber, or spray vinyl. These
types of surface materials, by their yielding, transform the line of contact
into a small area (surface) of contact and also may compensate for any
slight irregularities in the rolling surfaces.

Friction gearing is utilized in several forms of speed-controlling
devices. Possible combinations include:

1. Two equal cones
2. Cone and belt
3. Two equal rollers (cylinders)
4. Cylinder and hollow disk
5. Two equal hollow disks
6. Grooved or geared rollers, disks, and cylinders

SUMMARY

The computer generation of transmission paths were displayed from a
computer display program. The driving member of a computer-aided
mechanism was modeled in direct contact with the driven element.
These displays constituted the driver and follower and were programmed
to be in either pure rolling contact or sliding contact. Program segments
were used to illustrate the principles of interactive programming to display the lines of contact on both the driver and follower moving along a
display line common to both contact surfaces. During the course of this
chapter, several programming segments and basic modeling (FORTRAN
constructions) concerning two display techniques were presented. These

two techniques were presented from a software vs. firmware point of view.

The term computer modeling was used to define machine elements so that their shape, size, and turning axis could be stored in computer memory. These elements were then recalled to display a mechanism containing a driver and follower. The driver was turned about its axis and displayed together with the follower throughout its path of motion. Using the program segment approach, the designer was shown techniques for displaying unlimited types of drivers and followers. Even though a large host computer such as the IBM 370 Model 3033 was used together with the proper display screens, such as the Tektronix 618; the reader was not bogged down by long, detailed FORTRAN programs. In the case of sliding contact and speed ratios, enough programming was presented to describe the fundamentals of software-generated element displays. A short system-level display program was presented to recall a driver and a follower. This program example was followed by sample subroutines for a driver and follower, and the display of this software was shown in Figure 6.1.

Only the portion of the display program needed to generate Figure 6.2 was introduced. Program segments such as these help keep the user's interest on computer-aided machine design and avoid the pitfalls of computer programming for the sake of computer programming. How to display pitch points, angles of motion, and pressure angles was introduced so that the machine designer would be prepared for the computer-aided design of cams, gears, and other types of connectors.

A detailed description of the plotting of the conjugate to a given path was presented as related to the rotation of the driver and follower in rolling contact. Certain computer-aided principles were discussed and displayed in Figure 6.7 and followed by an interactive program segment. Other subroutines, such as SMOOT, were used to complete the rolling curves. Finally, machine elements in rolling contact, such as cylinders, cones, and friction gearing, were discussed in preparation for Chapter 7.

BIBLIOGRAPHY

Chasen, S. H., Geometric Principles and Procedures for Computer
    Graphic Applications. Prentice-Hall, Englewood Cliffs, N.J., 1978.
Giloi, W. K., Interactive Computer Graphics. Prentice-Hall,
    Englewood Cliffs, N.J., 1978.
Levinson, I. J., Machine Design. Reston Publishing, Reston, Va., 1978.
Parr, R. E., Principles of Mechanical Design. ' McGraw-Hill,
    New York, 1970.
Patton, W. J., Kinematics. Reston Publishing, Reston, Va., 1979.

Paul, B., Kinematics and Dynamics of Planar Machinery.
    Prentice-Hall, Englewood Cliffs, N.J., 1979.
Phelan, R. M., Fundamentals of Mechanical Design. McGraw-Hill,
    New York, 1970.
Ramous, A. J., Applied Kinematics. Prentice-Hall, Englewood Cliffs,
    N.J., 1972.
Ryan, D. L., Computer-Aided Graphics and Design. Marcel Dekker,
    New York, 1979.
Spotts, M. F., Design of Machine Elements. Prentice-Hall, Englewood
    Cliffs, N.J., 1978.

7

Gears and Cams

The design and computer display of gears and cams are often taken for granted because of their apparent simplicity. It was shown in Chapter 6 that one shaft could cause another to turn by means of two bodies in pure rolling contact. Cams are often designed on the basis of this principle. If the speed ratio must be exact or a rotary motion must be transferred as a rotation instead of a linear motion, toothed wheels called gears are used in place of a cam.

A cam can be displayed as a plate, cylinder, or other solid with a surface of contact so designed as to translate rotary motion to linear motion. The cam is mounted to the driving shaft, which rotates about a fixed axis. By the cam rotation, a follower is moved in a definite path. This is one of the applications of the principles discussed in Chapter 6. The follower may be a point (Chapter 1), a roller (Chapter 2), or a flat surface (Chapter 3). The follower may be attached to another part of the machine by a linkage (Chapter 5). The follower may move radially (Chapter 6) or the movement may be considered as that of a pair of gear teeth.

TYPES OF GEAR DISPLAYS

In Chapter 6 special attention was paid to the fact that rolling bodies may be used to connect axes that are parallel, intersecting, or neither. The same situations arise in the use of gears, and special names are given to the display of gears according to the situation for which they are designed. Gear displays may be so classified, as shown in Table 7.1. The term pinion used in Table 7.1 is often used to label the smaller of a pair of gears. The various other kinds of gear displays enumerated in Table 7.1 will be discussed in more detail after the terminology that applies to gearing in general has been considered.

Gear Terminology. The following display terms and their definitions will be easier to understand if Figure 7.1 is referred to occasionally.

Pitch diameter. The theoretical point of the contact between two gears as they mesh, also known as the pitch circle. It corresponds to the outside diameter of a cylinder that would transmit force or motion through a friction contact with another surface.

Diametral pitch. The ratio of the number of teeth to the number of display units in the pitch diameter.

Outside diameter. The diameter of the gear teeth at their outer edges.

Pressure angle. The angle displayed by the gear tooth contour and line of action from a line parallel or perpendicular to the center of the gear.

Circular pitch. The distance from the center of one tooth to the center of the next tooth measured on the pitch diameter of the gear.

Chordal (or tooth) thickness. The thickness of the tooth displayed at at the pitch diameter.

Addendum. The display distance from the pitch diameter of the gear tooth to its outer edge.

Dedendum. The display distance from the pitch diameter of the gear tooth to the bottom of the tooth form.

Whole depth. The total display height of the tooth equal to the addendum plus the dedendum.

Working depth. The depth the gear tooth extends into the open space between teeth when mating.

Face width. The length of the tooth from one side to the other.

Face of tooth. The surface area on the face width of the tooth between the pitch diameter line and the top of the tooth.

Flank of tooth. Similar to the display face except that it extends from the pitch diameter line to the bottom of the tooth. This includes the normal small fillet that is displayed at the bottom of the tooth.

Backlash. The space or clearance between the gear teeth as they are displayed at the pitch diameter line. This clearance is displayed on the back or nonworking face of the tooth. The amount of backlash will vary with the different gears shown in Table 7.1.

Fundamental Display Relationships

Machine designers require that all gear teeth be displayed with a slightly curved surface. This is known as an involute tooth form. This display form is necessary to assure ample clearance as the teeth mesh. If the teeth had straight sides, they would become damaged within a short time

Table 7.1   Types of Gear Displays

Axes connection	Type of gear	Classification of display
Parallel	External, internal, helical, herringbone, rack and pinion	Spur gear
Intersecting	Miter, crown, spiral	Bevel gear
Different plane	Hypoid, worm and wheel, helical (spiral)	Hyperboloidal (screw gears)

if a large amount of clearance were not provided. This display form is common to all types of gears, with the exception of worm gears.

The involute display contour, in addition to providing tooth clearance, also forms the tooth pressure angle. Gear teeth are usually displayed with two pressure angles: $14\text{-}1/2^{\circ}$ and 20°. A special 25° pressure angle is used for some of the gear types shown in Table 7.1. A pressure angle is displayed in all of the table types, including worm gears.

Another common display fundamental is that all gear teeth are spaced by diametral pitch. In a display program, DP is the storage location for the ratio of the number of teeth to the number of display units in the pitch diameter of the gear, and would be entered as

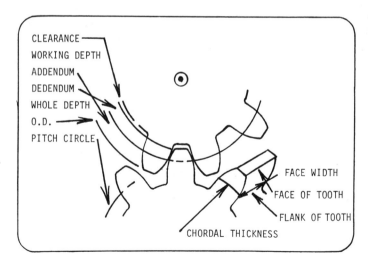

Figure 7.1   Gear display.

DP=NT/PD

where NT is the number of teeth and PD the pitch diameter. There is also a direct relationship between the DP of a gear tooth and its circular pitch, written in FORTRAN as

DP=3.1416/CP

where CP equals the circular pitch of the teeth. As a display example, the number of teeth on a gear, divided by the pitch diameter of the gear, would result in the display DP. Therefore, a modification using the outside gear diameter could also be used:

DP=(NT+2)/OD

where OD equals the outside diameter of the gear blank.

GEAR ELEMENTS AND DISPLAY TECHNIQUES

Figure 7.1 shows a partial pair of external spur gears in mesh with each other. Since these are the simplest form of gears, the following discussion will be based on this type of gear. Spur gears were the first gear type developed by machine designers. In this computer-aided approach to the design of machine elements, it is the next logical progression. It was shown in Chapter 6 that if two cylinders are keyed to their respective shafts, the angular speed of the first shaft changes as the angular speed of the second changes, provided that there is sufficient friction between the circumferences of the driver and follower. To make sure there will be no slipping (loss of friction), wheels having teeth around their circumferences are substituted for the smooth surfaces in Chapter 6.

Although there are currently many other displays for gears, spur gears are most common in machine design. It must be borne in mind, however, that the basic gear elements and display techniques are general and apply to the other types of gears shown in Table 7.1 as well as to spur gears. The advantages of spur gears include ease of manufacture, reliability, minimum redesign, and ease of design assembly. Because of their design simplicity, spur gears are made of many types of metallic and nonmetallic materials.

Their straight or parallel tooth design enables the spur gear to be made in large diameters and broken down or split for assembly as segments in a field application without requiring special machinery design. In addition, the straight tooth design eliminates all end thrust or side movement of the gears compared to gears having tapered tooth faces.

Computation of Speed Ratios

The profiles of spur gear teeth must be such that the speed ratio is constant. A computer file must be generated for each tooth displayed.

The file contains the plotting routines for straight tooth design. These straight tooth designs are simple to display on an output device, but because of their simplicity, they have some design disadvantages. Because of the minimum number of teeth in contact between the pinion and driven gears, they are usually designed only for slow or medium-speed machinery. As shown in Figure 7.2, spur gears are displayed as regular gears and also as straight gears called <u>racks</u>. The racks are normally designed to transmit rotary motion from the driving pinion into horizontal or linear motion of an object.

Now that the profile for each of the common DPs can be generated by

CALL TOOTH(X,Y,DP,THETA)

where CALL TOOTH is the name given to the computer subroutine for location in storage and the list of arguments are:

X = horizontal screen location for lower left corner of profile
Y = vertical screen location for lower left corner of profile
DP = diametral pitch of the spur gear tooth
THETA = rotation of tooth around gear blank

The designer may display working spur gears. Such a pair is shown in Figure 7.3. Here the larger gear has 16 teeth and the smaller 12. Assume that the designer did not have CALL GEAR(X,Y,NT,OD) stored in computer memory and that shaft S is being animated from computer memory. The 16 tooth gear, labeled A, will turn with shaft S. As the gear blank turns through $(360/16)^{\circ}$ segments, or 22.5° for each THETA; the display

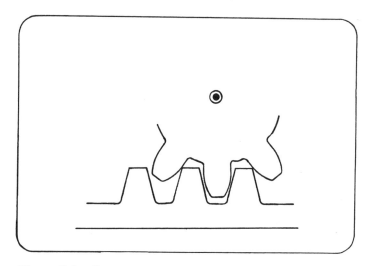

Figure 7.2   Spur gears.

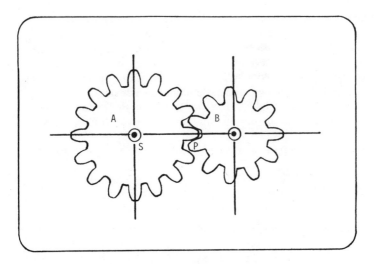

Figure 7.3   Display of gear program.

program calls the subroutine TOOTH. Each tooth of gear A is displayed
as it rotates on shaft S. If the center locations of gear blank A and B are
properly placed in the display program, the teeth of gear blank B may be
displayed in the same manner. As gear blank B is pushed by gear tooth A
through $(360/12)^\circ$ or 30° segments; the display program calls the subroutine TOOTH to display each of the teeth on gear B. As the program is
executed, the viewer notes that the teeth on A will push the teeth on B, a
tooth on A coming in contact with a tooth on B and pushing that tooth along
until the gears have turned so far around that those two teeth swing out of
reach of each other or come out of contact. But before these two teeth
come out of contact, another pair of teeth must come in contact, so that
gear A will continue to drive gear B. For B to make a complete revolution, each of its 12 teeth must be pushed past the center line. Therefore,
while B turns once, gear A turns 12/16 of a turn, since A has 16 teeth in
all. It is evident from viewing the display that in order for the teeth to
mesh, the distance from the center of one tooth to the center of the next
tooth on both A and B must be alike.

The point at which gear A pushes gear B is labeled P in Figure 7.3.
Through this point the programmer may display circles about SA and SB
as centers. These circle diameters are stored as DA and DB. Then the
distance between gear centers is

$$CD = DA/2.+DB/2.$$

where the two gears when turning will have the same speed ratio as two
rolling cylinders of diameters DA and DB. The point P that divides the

line of centers of the two gears is called the pitch point. The display
circles are called pitch circles for gear A and B.

The distance from the center of one CALL TOOTH to the center of
another CALL TOOTH, measured on the pitch circle, is called the
circular pitch. This may be computed from the display program as

PC = 3.1416*DA/NT

Display of Gear Elements

The display subroutine CALL TOOTH is the software ordinarily used to
designate the tooth size, location, and rotation. It is equal to the number
of teeth divided by the diameter of the pitch circle. Often in designating
the size of a computer-generated gear the word "pitch" is used without the
adjective "diametral." For this reason, the diametral pitch is sometimes
called the pitch number. The diametral pitch is expressed by the FORTRAN
statement

DP=NT/PCD

where DP is the diametral pitch or pitch number, NT the number of teeth,
and PCD the pitch circle diameter. Two types of teeth subroutines may be
used, as shown in Figure 7.4.
With the subroutines displayed in Figure 7.4 and the additional subroutines
shown in Table 7.2, a complete gear representation can be programmed.

The display curves that form the profile of the subroutines for gear
teeth may, in theory, have any form whatever, provided that the profiles
conform to the following fundamental machine design law:

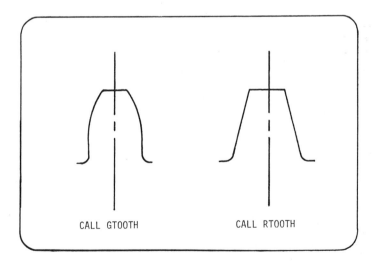

CALL GTOOTH   CALL RTOOTH

Figure 7.4   CALL tooth displays.

Table 7.2   List of Gear Elements

Subroutine name	Argument list
ADDEND	XSCRN, YSCRN, OD
DEDENT	XSCRN, YSCRN, RC
TDEPTH	ADDEND + DEDEND
CLEARA	DEDEND - ADDEND
TTHICK	PC/NT
SPACER	OD/NT
BACKLH	SPACER - TTHICK
PANGLE	THETA (14-1/2,20,25)

A line displayed from the pitch point where the teeth are in
contact must be perpendicular to a line displayed through
another point of contact tangent to the curves of the teeth,
or the common normal to the tooth curves at all points of
contact must pass through the pitch point.

This was illustrated in Figure 7.3. The teeth on the center-line position
touch each other at P; that is, the curves of the CALL TOOTH are tangent
to each other at this point. If a line is displayed tangent to the two curves
at P, and the angle measured between this line and the center line, the
viewer will see the angle of the path of contact. The actual path of contact is a line displayed through all the points at which the teeth of a gear
touch each other during the display program. This path may be straight
or curved, depending upon the nature of the curves that form the subroutine TOOTH. For every different position that CALL TOOTH occupies
during the display of one pair of teeth, the curves have a different point of
contact. However, in all properly constructed gear displays, the pitch
point P is one point on the path of contact.

The angle of the path of contact is not the pressure angle. The
pressure angle is the angle between a line displayed through the pitch
point P perpendicular to the line of centers and the line displayed from
the pitch point P to a point where a pair of teeth are displayed on contact.
In some forms of gear teeth display, this angle remains constant; in
others, it varies. This is because the direction of the force that the
driving tooth exerts on the driven tooth is always along the line displayed
from the pitch point to the point where a pair of teeth are displayed in
contact. The smaller the pressure angle, the greater will be the component of the force in the direction to cause the driven gear to turn, and

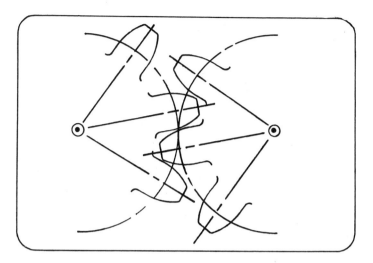

Figure 7.5   Displaying gear teeth.

the less will be the tendency to force the shafts apart. In other words, a large pressure angle tends to produce a large pressure on the shafts.

Displaying Gear Teeth

When the tooth subroutines for teeth outlines have been written and the circular pitch, backlash, addendum, and clearance arguments have been added to the display program, the teeth may be output as shown in Figure 7.5. Let CALL GTOOTH be the subroutine outline for gears 1 and 2. The program will display three teeth on each gear, one pair of which will be in contact at the pitch point. The following program steps are then taken:

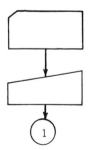

1. The proper JC L (job control language) for the subroutines listed in Table 7.2 is used, plus PLOT-10.
2. CALL INITT(240) is used for output to a direct-view storage tube (DVST) device with a graphics tablet attachment.

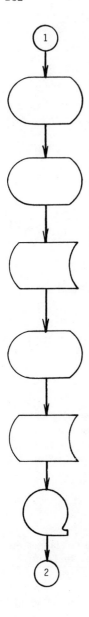

3. Display the addendum circle of each
    gear with a radius of pitch circle
    plus addendum.
4. Display the addendum circle of each
    with a radius equal to pitch circle
    minus an amount equal to the adden-
    dum of mating gear plus the clear-
    ance.
5. Space off the circular pitch on either
    side of P on each pitch circle. This
    may be done using the graphics tablet
    and an electric pencil. Start with a
    line tangent to the pitch circles at P
    and locate each of the X, Y starting
    positions for CALL GTOOTH.
6. Display each of the starting locations
    and enter each rotation (THETA) from
    the keyboard into the program storage
    matrix.
7. From the graphics tablet, enter the
    necessary scalar lines to represent
    pressure angle, line of centers, and
    pitch circles.
8. From the program storage matrix,
    supply the argument lists for each of
    the six teeth to be displayed on the
    DVST.

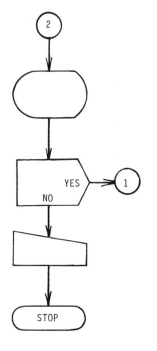

9. Display the gear teeth as selected in program segment 5.
10. Program modifications may be made prior to the production of hard copies or plotter drawings.
11. CALL FINITT (0, 0) is used to clear the DVST of further graphical information.
12. The program stops.

By viewing the process described above, the designer checks for such things as tooth curves that have flanks that extend into the root line. This will result in a weak tooth and the designer avoids this by placing a small fillet at this junction. The size of this fillet is limited by the arc of a circle connecting the root line with the flank and lying outside the actual path of the end of the face of the other gear tooth (mating). This path of the end of the face is called the true clearing curve. This curve is the path traced by the outermost corner of one tooth on the plane of the other gear. This is always known from information given in programming segments 5 and 6. The path can be displayed in segment 10 for easy reference of the designer.

GEAR STUDIES AND COMPUTER ANALYSIS

The form of the gear display most commonly given to computer analysis is that known as the involute of a circle. Gear teeth constructed with this curve will conform to the fundamental law of machine design discussed earlier. This display curve and methods of producing it will, therefore, be studied before the method of applying it to gear teeth is discussed.

A typical involute curve may be displayed by placing a jar lid on the graphics tablet. Attached to the edge of the jar lid is a very fine wire which is wrapped around the electric pencil. If the operator keeps the wire taut and begins to wrap the wire around the jar lid, causing the wire to become shorter as the pencil traces a curve on the graphics tablet; an involute of the circle represented by the jar lid is displayed. The same result is obtained from a FORTRAN program segment for displaying the tracing point to be carried by a line rolling on a circle discussed in Chapter 6. Using either method, all involutes from the same circle are alike, but involutes displayed from circles of different diameters are different. The greater the diameter of the circle, the flatter will be its displayed involute.

In considering the involute of a circle to be used in automated gear tooth display, it is, of course, impractical to stop and wrap a wire around a jar lid. Instead, Figure 7.6 shows the method of displaying an involute for the subroutine TOOTH. Suppose that the involute is to be displayed starting from any point P on the circle whose center is C. The subroutine selects any convenient distance (usually one-eighth the diameter of the circle). This distance is then located (placed from P in a clockwise direction) as distant points i, j, k, l, and so on. At each of these distant points a perpendicular scalar is computed through point C. Each of these perpendicular scalars will mate with scalars that are tangent to each of the points i, j, k, and l. The subroutine now calculates new distances along these tangent scalars based upon the following relationships:

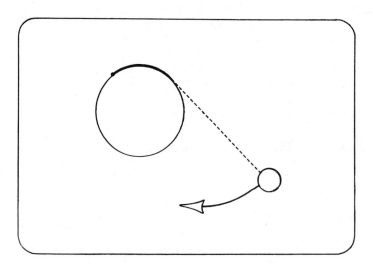

Figure 7.6   Display of involute.

1. On the tangent to i locate a distance of 1*Pi
2. On the tangent to j locate a distance of 2*Pi
3. On the tangent to k locate a distance of 3*Pi
4. On the tangent to l locate a distance of 4*Pi

These new point locations are then connected through a CALL SMOOT
and displayed as a smooth curve that will be the true involute.

Involute Spur Gears

Involute spur gears can be arranged in a number of different patterns.
Figure 7.7 is a display of a pinion (driver) and rack in mesh. No new
principles are involved since the rack is merely a spur gear whose
radius of the pitch circle has become infinite. The base line of the
rack is tangent to the pressure angle at infinity. For this reason the
sides of the teeth of the rack will be straight lines perpendicular to
the line formed by the pressure angle. In Figure 7.7 the addendum on
the rack is displayed in mesh as much as the pinion will allow. The
addendum of the pinion will display the end of the path of contact.

In Figure 7.8 a pinion is displayed inside an annular gear. The
addendum of the annular is displayed by the tangent point of the pinion
base circle and the pressure angle. The addendum of the pinion is not
limited except by the DP of each tooth becoming pointed. The annular
gear may be though of as a rack that has crossed the infinite radius
point and makes its radius a finite negative quantity. In other words,
the base circle of the annular gear lies inside the path of contact.
For example, if some point on the pressure angle is chosen and the

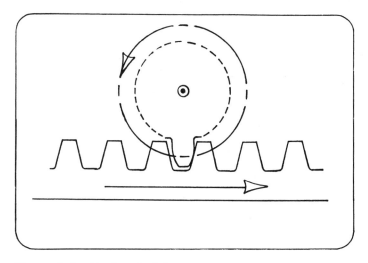

Figure 7.7   Rack and pinion.

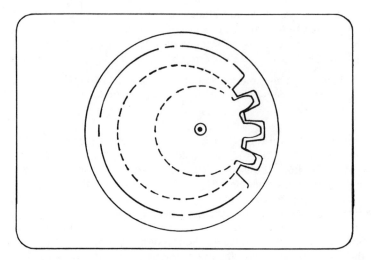

Figure 7.8   Annular and pinion.

tooth curves displayed as they would appear in contact at that point,
the teeth profiles of the annular will be found to be concave, and the
addendum of the annular is limited by the base circle.

Cycloidal Gears

Cycloidal gear teeth are designed by a software system in which the
faces of the teeth are computed as epicycloids generated on the pitch
circles. The flanks of the teeth are then hypocycloids generated inside
the pitch circles. Few, if any, spur gears are designed with this
analysis in mind. It is best used in certain types of bevel or spiral
gears. The best known use is in the case of high-tolerance gearing,
where interchangeability can be a problem. For these cases the
cycloidal system is ideal because the same describing circle can
and must be used in generating all the faces and flanks. The size
of the describing circle depends on the properties of the hypocycloid
curve which forms the flanks of the teeth. If the diameter of the
describing circle is programmed to be half of the pitch circle, the
flanks will be radial, which displays a comparatively weak tooth at
the root. If the describing circle is entered smaller, the hypocycloid
curves away from the radius and will display a strong form of the tooth.
However, if the describing circle is input too large, the hypocycloid
will curve the other way, passing inside the radial lines and displaying
a very weak tooth.
    From displays of the system we have been discussing, a practical
conclusion would appear to be that the diameter of the describing circle
should not be more than one-half that of the pitch circle of the smallest

gear of a pair. Two versions of the display system have been used,
one with radial flanks on a 12-tooth gear and one with radial flanks on
a 15-tooth gear. The latter is usually referred to as the standard
interchangeable series.

Hypoid

Hypoid gears are a special design resulting from an effort to obtain
satisfactory gears for connecting nonparallel and nonintersecting shafts.
The appearance of hypoid gears are similar to cyloidal bevel gears. The
teeth on both types are generated by the use of rotary translation
(epicycloids and helix). The pinion of a pair of hypoid gears is larger
than the pinion of a pair of cycloidal gears with the same number of
teeth. This is responsible for the pinion of hypoid gears being stronger
than that of cycloidal gears. Hypoid gears, because of the helix translation, have a continuous pitch line contact and a larger number of teeth
in contact than cycloidal gears. Hypoid gears are quieter and wear
longer than cycloidal gears; for this reason they are used in the differ-
entials of automobiles to lower the drive shaft. Hypoid gears are used
in machine design where nonintersecting and nonparallel shafts or over-
lapping shafts are required.
     From the three major computer-generated gear teeth—involute,
cycloidal, and hypoid—several applications can be designed and displayed:

1. External spur
2. Rack and pinion
3. Annular and pinion
4. Mated interchangeable high-tolerance
5. Stepped
6. Twisted spur or helical
7. Herringbone
8. Pin
9. Bevel (crown, internal, spiral)
10. Skew (hyperboloidal gears, such as worm and wheel or helical)

TYPES OF CAM DISPLAYS

A cam and its follower form an application of the principle of transmitting
motion by direct sliding contact, as presented in Chapter 6. As in the
case for gears, various situations arise for the use of cams, and special
names are given to the display of cams according to the situation for
which they are designed. Cam displays classified on this basis are listed
in Table 7.3.
     Figure 7.9 is a display of a plate cam, and Figure 7.10 is a display
of a cylindrical cam. Many machine designs depend largely upon cams to
give motion to the various parts. Nearly all cams are designed for the
special purpose intended. This being the case, speed ratios are not the

Table 7.3   Types of CAM Displays

Motion of follower	Type of CAM	Display
Perpendicular	Plate	Single-view
Offset perpendicular	Plate	Single-view
Radial perpendicular	Plate or cylinder	Two-view
Parallel	Cylinder ·	Two-view
Offset parallel	Cylinder	Displacement diagram
Radial parallel	Cylinder	Follower analysis
Intermittent	Triangular	Single-view
Multiaction	Double-grooved	Animation
Lever-action	Combination	Simulation

desired output; rather, a cam assigns a certain series of definite positions that the follower is to assume while the driver occupies the corresponding series of positions.

The graphic relationship between the successive positions of the driver and follower in a cam motion are displayed as a displacement diagram. Displacement diagrams are part of the design program discussed later in the chapter. For now it is enough to know that the abscissas of displacement diagrams are linear distances arbitrarily

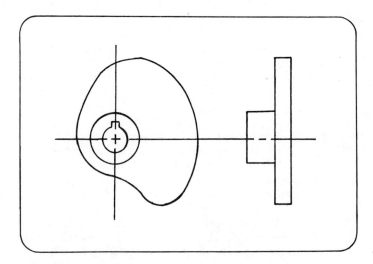

Figure 7.9   Plate cam.

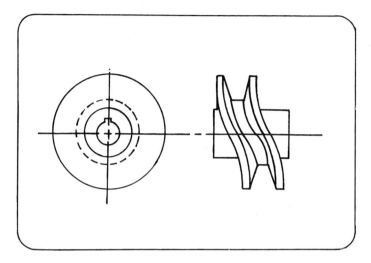

Figure 7.10   Cylindrical cam.

chosen to represent angular motion of the cam and the ordinates are the corresponding displacements of the follower from its initial position. This is illustrated in Figure 7.11, where the line OABCD represents the motion given by the cam. The perpendicular distance of any point in the line from the axis 0Y represents the angular motion of the driver. The perpendicular distance of the point from the 0X represents the corresponding movement of the follower, from some point considered as a

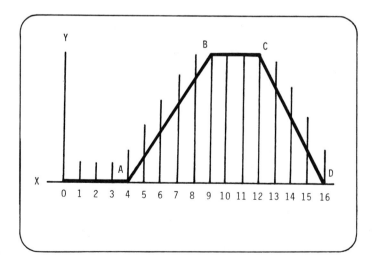

Figure 7.11   Displacement diagram.

starting point. Thus, the line of motion 0ABCD indicates that from
driver positions 0 to 4, the follower had no motion; from driver positions 4 to 9, the follower had a uniform upward motion; from driver
positions 9 to 12, the follower had no motion; and from driver positions
12 to 16, the follower had a uniform motion downward, returning it
again to its starting point.

Follower Motions

A machine design often requires that a cam transmit a definite displacement to the follower in a short interval of time, the nature of the motion
not being programmed. For the displacement diagram shown in Figure
7.11, the follower has two uniform motions and two periods of rest or
dwell. If the cam is programmed to revolve quickly, quite a shock will
occur where the motion changes. To soften this, the form of the diagram
can be modified, provided that it is allowable to change the nature of the
motion.

Suppose that the cam was to raise the follower rapidly from A to B
in Figure 7.11, the nature of the motion to be such that the shock will be
as light as possible. For the straight, flat line 0A, a dwell is present.
The motion changes suddenly at A, accompanied by a perceptible shock.
This could be modified by the addition of a function to replace the linear
(uniform) motion represented by A-B. It is often possible to have the
motion of the follower guided by some definite law, such as uniform,
harmonic, or parabolic. The laws of these motions were developed in
Chapter 2 and can now be used when displaying displacement diagrams.

Plate, Cylindrical, and Combination

A cam imparts motion to a follower guided so that it is constrained to
move in a plane that is perpendicular to the axis about which the cam rotates. This type of cam is best displayed as a flat plate. Cams may also
occupy a plane coincident with or parallel to a plane in which the cam
rotates. This type of cam is best displayed as a cylinder and follower.
The displacement transmitted to the follower depends upon the shape of
the cam. The follower may move continuously or intermittently; it may
be displayed with uniform speed or variable speed; or it may have
uniform speed part of the time and variable speed part of the time. A
knowledge of the various types of cams, and an idea of the manner of
attacking the design program for output of a suitable cam for any specific
purpose, can best be obtained by studying a programming method for a
number of examples.

1. The proper JCL (job control language)
    for the cam types listed in Table 7.3
    is used, plus PLOT-10.

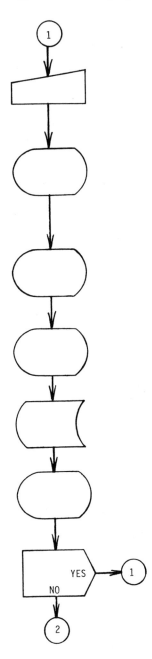

2. CALL INITT(240) is used for output to
    a DVST device with a graphics tablet
    attachment.
3. Display the base circle for locating the
    relative positions of the camshaft axis
    and the follower. The base circle is
    displayed at the center of the cam-
    shaft with a radius equal to the least
    distance between that center and the
    pitch profile.
4. Display the pitch profile or pitch line
    traveled by the reference point on the
    follower during one revolution of the
    cam. The reference point is the center
    of a roller follower and the lowest
    touching point of contact in all other
    followers.
5. Display the cam shaft in cross section.
6. Using the graphics tablet, enter the
    devices for holding the follower in
    place together with the type of follower
    desired.
7. Display the displacement diagram for
    the follower selected above.
8. Program modifications may now be
    made before hard copies or plotter
    drawings are produced.

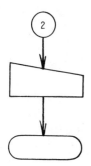

9. CALL FINITT (0, 0) is used to clear
the DVST of further graphical information.
10. The program stops.

In various computer-aided design machines the movements of parts
that have to be timed with respect to each other are often programmed for
two or more cams properly designed and properly displayed to give each
piece its desired motion at the required time. For example, a cylindrical
cam and a plate cam might be arranged to work in combination. The
cylindrical cam might make two or more revolutions for every one of the
plate cam.

If the shape of the cam is known but nothing is known about the
character of the motion of the follower, an analysis program may be
used to display the cam in a series of positions, computing the corresponding displacements of the follower, displaying or plotting a displacement-
time curve as explained in Chapter 2, and then obtaining the velocity and
acceleration from this curve. In some programs the velocity curve may
be obtained by velocity vector displays, as explained in Chapter 3; other
programs use acceleration vector displays (Chapter 4).

DESIGN PROGRAMS

```
C    ****************************************************************
C    *                                                              *
C    * BEGINNING OF GRAPHIC LIST ....... DESIGN PROGRAM LIST-       *
C    * ING TO FOLLOW                                                *
C    *                                                              *
C    ****************************************************************
C
C
C
C    BBBBBB  EEEEE  GGGGG  II  NN    NN  LL        II  SSSSSS  TTTTT
C    BB  BB  EE     GG     II  NNN  NN   LL        II  SS         TT
C    BBBBB   EEEEE  GG GGG II  NNNNNN    LL        II  SSSSSS      TT
C    BB  BB  EE     GG  GG II  NN    NN  LL        II      SS      TT
C    BBBBBB  EEEEEE GGGGG  II  NN    NN  LLLLLL II  SSSSSS      TT
C
C
```

```
C      CAMPLOT IS A SOFTWARE PACKAGE TESTED UNDER WATFIVE
C      & FORTRAN IV G1 COMPILERS ON AN IBM 370/3033 USING A
C      CALCOMP PLOTTER AND THE CALCOMP VENDOR SOFTWARE.
C      NOTE:  NOT ALL SUBROUTINES HAVE BEEN CONVERTED TO
C      PLOT-10 AND ARE NOT FUNCTIONAL ON A DVST.
C
C
C      ******************************************************************
C      *                                                                *
C      * INDEX OF SOFTWARE CONTENTS FOR CAMPLOT DESIGN      *
C      * PROGRAMS ....                                                   *
C      *                                                                *
C      ******************************************************************
C
C
C
C      1)  INDEX
C
C      2)  HEADING SUBROUTINES
C
C            A    BEGIN
C            B    TITLE
C                  I    FILLIT
C                  II   MESAGE
C                  III  ENDD
C
C      3)  CAM PROFILE SUBROUTINES
C
C            A    PLATE
C            B    CYLIND
C            C    TRIANG
C
C      4)  CONSTRUCTION (DRAFTING) SUBROUTINES
C
C
C            A    ANGSP
C            B    CENTER
C            C    CIRCLE
C            D    DASH
C            E    DASHCR
C            F    DIMLIN
C            G    POOCHE
C            H    SLINE
C            I    LEADER
C            J    WIRE
C
```

```
C     5) CAM PART FEATURE SUBROUTINES
C
C           A    TPHOLE
C           B    ELHOLE
C           C    SHAFTS
C
C
C     6) MATHEMATICAL FUNCTION SUBROUTINES
C
C           A    AXDIST
C           B    ARTAN
C           C    CCOORD
C           D    DSCRT
C           E    PCOORD
C           G    SIMON
C
C
C     **************************************************************
C     *                                                            *
C     * HEADING SUBROUTINES ... HEADING SUBROUTINES ... HEAD-      *
C     * ING SUBS                                                   *
C     *                                                            *
C     **************************************************************
C
          SUBROUTINE BEGIN
C
C     THIS SUBROUTINE INITIATES CALL INITT(240) FOR THE DVST
C     TERMINAL AND CALL PLOTS FOR THE CALCOMP PACKAGE.
C     THE CONVERSION SUBROUTINE PLOT USED IN PLACE OF
C     CALCOMP COMMANDS FOR GRAPHIC ROUTINES IS LOCATED
C     IN THE 370 AS "TEKPLOT".
C
C
          CALL PLOTS
          CALL INITT(240)
          CALL TEKPLOT
          RETURN
          END
C
C
          SUBROUTINE TITLE
C
C     THIS SUBROUTINE IS A STANDARD TITLE BLOCK FROM THE
C     PUBLICATION "COMPUTER AIDED GRAPHICS AND DESIGN"
C     WRITTEN BY DANIEL L. RYAN AND PUBLISHED BY MARCEL
C     DEKKER INC.
C
```

```
             CALL TTITLE
             RETURN
             END
C
C        SUBROUTINE FILLIT(TX,TY)
C
C
C        THIS SUBROUTINE FILLS THE CAM TITLE SPACES WITH THE
C        CAM NUMBER, DATE, SCALE, AND RELATED LABELS.
C        "FILLIT" CALLS SUBROUTINE MESAGE, WHICH READS THE
C        RESPONSE TO A PROMPT AND CALLS SUBROUTINE SYMBOL,
C        WHICH ENTERS THE DESIRED MESSAGES.
C
             DIMENSION MESS(74)
             WRITE(3,9999)
9999         FORMAT(///,'**GRAPHIC EXEC READ UNDER SUB FILLIT**')
             WRITE(3,1000)
1000         FORMAT(' DESIGNED BY')
             CALL MESAGE(MESS,ICT)
             WRITE(3,2000)
2000         FORMAT(' THE CAM NUMBER IS')
             CALL MESAGE(MESS,ICT)
             WRITE(3,3000)
3000         FORMAT(' THE DRAWING DATE IS')
             CALL MESAGE(MESS,ICT)
             WRITE(3,4000)
4000         FORMAT(' SCALE IS')
             CALL MESAGE(MESS,ICT)
             WRITE(3,5000)
5000         FORMAT(' THE DRAWING TITLE IS')
             CALL MESAGE(MESS,ICT)
             RETURN
             END
C
C
C   ***************************************************************
C   *                                                             *
C   * CAM PROFILE SUBROUTINES...CAM PROFILE SUBROUTINES...*
C   * CAM                                                         *
C   *                                                             *
C   ***************************************************************
C
C
C
```

```
      SUBROUTINE PLATE
C
C
C     THIS SUBROUTINE USES TWO NESTED ROUTINES FOR DISPLAYING
C     THE PLATE PROFILE AND THE CORRESPONDING DISPLACEMENT
C     DIAGRAM.  SUBROUTINE PROFIL DISPLAYS THE CAM OUTLINE,
C     WHILE SUBROUTINE DISDIA WILL PLOT THE FOLLOWER MOVE-
C     MENT IN A TYPICAL DISPLACEMENT DIAGRAM.
C
      REAL TA1(6),TA2(6),TA3(6)
      READ(3,100) TA1,TA2,TA3
100   FORMAT(18A4)
      CALL PROFIL
      CALL DISDIA
      RETURN
      END
C
C
C
C
C
C
      SUBROUTINE PROFIL
      DIMENSION XDOT(50),YDOT(50),XA1(4),YA1(4)
      READ(3,200) XBGN,YBGN,YLGN,NDIV
200   FORMAT(4(F6.3,1X),I3)
      XA1(3)=0.0
      YA1(3)=0.0
      XA1(4)=1.0
      YA1(4)=1.0
      XDIV=NDIV
      XEND=XBGN+XLGN
      YEND=YBGN+YLGN
      CRAD=(YEND-YBGN)/2.
      CALL PLOT(XBGN,YBGN,3)
      CALL PLOT(XEND,YBGN,2)
      CALL PLOT(XEND,YEND,2)
      CALL PLOT(XBGN,YEND,2)
      CALL PLOT(XBGN,YBGN,2)
      XDOT(1)=XBGN
      XDOT(NDIV+1)=XEND
      DO 300 I =2,NDIV
      XDOT(I)+XDOT(I-1)+XINC
      CALL PLOT(XDOT(I),YBGN,3)
      CALL PLOT(XDOT(I),YEND,2)
```

```
300       CALL PLOT(XDOT(I),YBGN,3)
          XNOW=XBGN
          CDIV=180.0/XDIV
          CALL CIRCL(XNOW,YBGN,270.0,90.0,CRAD,CRAD,0.0)
          YCTR=YBGN+CRAD
          YNOW=YBGN
          SANG=270.0
          EANG=90.0
          ND0=NDIV
          D0 400 I=2,NDC
          EANG=SANG-CDIV
          CALL CIRCL(XNOW,YNOW,SANG,EANG,CRAD,CRAD,0.0)
          SANG=EANG
          CALL WHERE(XCIR,YCIR,FCTR)
          XA1(1)=XCIR
          XA1(2)=XDOT(I)
          YA1(1)=YCIR
          YA1(2)=YCIR
          CALL PLOT(XBGN,YCIR,2)
          CALL PLOT(XCIR,YCIR,3)
          CALL DASH(XA1,YA1,2,1)
          CALL PLOT(XCIR,YCIR,3)
          XNOW=YCIR
          YNOW=YCIR
400       YDOT(I)=YCIR
          YDOT(1)=YBGN
          NDIV=NDIV+1
          YDOT(NDIV)=YEND
          XDOT(NDIV+1)=0.0
          YDOT(NDIV+2)=0.0
          XDOT(NDIV+2)=1.0
          YDOT(NDIV+2)=1.0
          CALL FLINE(XDOT,YDOT,-NDIV,1,0)
          RETURN
          END
C
C
C
          SUBROUTINE DISDIA
          REAL TA1(6),TA2(6),TA3(6)
          READ(3,500) TA1,TA2,TA3
500       FORMAT(6A4)
          CALL PLOT(0.0,2.55,-3)
          XHOL=1.125
          YHOL=.225
          D0 600 I=1,3
          CALL CIRCL(XHOL,YHOL,180.0,-180.0,0.125,0.125,0.0)
```

```
600       XHOL=YHOL+4.250
          CALL RECT(0.,0.,11l.,8.5,0.,3)
          CALL RECT(0.005,0.005,11.,8.5,0.,3)
          CALL PLOT(8.,0.,3)
          CALL PLOT(8.,1.,2)
          CALL PLOT(11.,1.,2)
          CALL PLOT(8.,.6,3)
          CALL PLOT(11.,.6,2)
          CALL PLOT(8.,.3,3)
          CALL PLOT(11.,.3,2)
          CALL PLOT(0.,8.,3)
          CALL PLOT(11.,8.,2)
          CALL SYMBOL(8.1,.605,.12,TA1,0.,24)
          CALL SYMBOL(8.1,.305,.1.TA2,0.,24)
          CALL SYMBOL(8.1,.005,.1,TA3,0.,24)
          RETURN
          END


C
C
C
C         SUBROUTINE CYLIND
C
C
C         CYLIND DISPLAYS THE THREE STANDARD ORTHOGRAPHIC
C         PROJECTIONS AND AN OPTION TO DISPLAY A PICTORIAL
C         CAM.  IF ANG=0, THE PICTORIAL IS SKIPPED.
          COMMON /CUPID1/ X(100),Y(100),Z(100) IP(100),NDATA
C
          CALL RDC1 (NDATA,XSET,YSET,X0,Y0,ANG,SF,SF1,X,Y,Z,IP)
          CALL FACTOR(SF)
          CALL PLOT(X0,Y0,-3)
          IF(ANG.NE.0.) CALL IS0(ANG,SF1)
          CALL FVIEW
          CALL TVIEW(YSET)
          CALL PVIEW(XSET)
          CALL FACTOR(1.)
          RETURN
          END
C
C
C
```

```
            SUBROUTINE IS0(ANG,SF)
            COMMON /CUPID1/X(100),Y(100),Z(100),IP(100),NDATA
            DIMENSION XPLOT(100),YPLOT(100)
            CALL FACTOR(SF)
            COSA=COS(ANG/57.3)
            SINA=SIN(ANG/57.3)
            D0 700 I=1,NDATA
            XPLOT(I)=(X(I)+Z(I)*COSA)+12.5
            YPLOT(I)=(Y(I)+Z(I)*SINA)+6.67
700         CALL PLOT(XPLOT(I),YPLOT(I),IP(I))
            CALL FACTOR(1.0)
            RETURN
            END
C
C
C
            SUBROUTINE FVIEW
            COMMON /CUPID1/ X(100),Y(100),Z(100),IP(100),NDATA
            D0 800 I=1,NDATA
800         CALL PLOT(X(I),Y(I),IP(I))
            RETURN
            END
C
C
C
            SUBROUTINE TVIEW
            COMMON /CUPID1/ X(100),Y(100),Z(100),IP(100),NDATA
            DIMENSION Z1(100)
C
C
            D0 900I=1,NDATA
            Z1(I)=Z(I)+YSET
900         CALL PLOT(X(I),Z1(I),IP(I))
            RETURN
            END
C
C
C
            SUBROUTINE PVIEW
            COMMON /CUPID1/ X(100),Y(100),Z(100),IP(100),NDATA
            DIMENSION Z1(100)
            D0 950 I=1,NDATA
            Z1(I)=Z(I)+XSET
950         CALL PLOT(Z1(I),Y(I),IP(I))
            RETURN
            END
C
```

```
C
C
          SUBROUTINE TRIANG
C
C     TRIANG IS A CAM ANIMATION PROGRAM FOR DISPLAYING EITHER
C     PLATE OR CYLINDRICAL CAMS IN MOTION ON THE DVST SCREEN.
C     THIS SUBROUTINE MAKES USE OF THE CONSTRUCTION ROUTINES
C     LISTED IN THE INDEX PORTION OF THIS SOFTWARE LISTING.
C
          DIMENSION P(100,3),IC(200),VP(100),PP(100,3)
C
C     P IS AN ARRAY OF THREE DIMENSIONAL DATABASE IN WIREFORM.
C
C     IC IS AN ARRAY CONTAINING THE POINT CONNECTIONS FROM
C     ARRAY P.
C
C     VP IS AN ARRAY CONTAINING AMOUNTS OF ROTATION IN X, Y,
C     AND Z.
C
C     PP IN AN EMPTY ARRAY TO BE USED FOR TRANSFORMED
C     DATABASE.
          COMMON /CUPID2/ NP,NC,NV,SPACE,P(I,J),IC(NC),
          VP(NV,3) CALL FACTOR(SF)
          CALL RDC2(P,NP,VP,NV,IC,NC,SPACE)
C
C     EACH BUFFERED FRAME HAS THE FOLLOWING TRANSFORMATIONS
C     AND DISPLAYS:
C
          D0 101 I=1,NV
C     CONVERT 3-D X,Z ROTATIONS FOR TRANSFORMATION TO THE
C     XSCRN LOCATION
C
          A=ARTAN(VP(I,1),VP(I,3))
          SA=SIN(A)
          CA=COS(A)
C
C
C     CONVERT 3-D X POINTS FOR SCREEN COORDINATES.  PREPARE
C     3-D Z POINTS FOR FINAL ROTATION TRANSFORMATION TO DVST.
C
          D0 201 J=1,NP
          PP(J,3)=P(J,3)*CA+P(J,1)*SA
          PP(J,1)=P(J,1)*CA-P(J,3)*SA
 201      CONTINUE
```

```
C
C     CONVERT Y ROTATION FOR Y TRANSFORMATION TO DVST.
C
      VPP=VP(I,3)*CA+VP(I,1)*SA
      A=ARTAN(VP(I,2),VPP)
      SA=SIN(A)
      CA=COS(A)
C
C     CONVERT 3-D Y, Z COORDINATES TO DVST Y.
C
      DO 301 J=1,NP
      PP(J,2)+P(J,2)*CA-PP(J,3)*SA
 301  CONTINUE
C
C     ADD X, Y TRANSFORMATIONS TO THE POINTS FOR PLOTTING
C     ANIMATION.
C
      DO 401 K=1,NP
      PP(K,2)=PP(K,2)+6.0
      PP(K,1)=PP(K,1)+SPACE*I
 401  CONTINUE
C
C     FOR EACH BUFFERED FRAME, POINTS ARE CONNECTED WITH
C     DARK VECTORS FOR NEGATIVE ELEMENTS IN THE ARRAY;
C     BRIGHT VECTORS ARE DISPLAYED WITH POSITIVE ELEMENTS
C     IN THE ARRAY.
C
      DO 501 J=1,NC
      IF(IC(J).LT.0)GOTO901
      CALL PLOT(PP(IC(J),1),1),PP(IC(J),2),2)
      GOTO 501
 901  K=-IC(J)
      CALL PLOT(PP(PP(K,1),PP(K,2),3)
 501  CONTINUE
 101  CONTINUE
C
C     MOVE BUFFERED FRAME TO NEW LOCATION FOR DVST VIEWING.
C
      XSET=(SPACE*NV)+8.0
      CALL FINITT(XSET,0.)
      CALL FACTOR
      RETURN
      END
C
C
C
```

SUMMARY

The designs and computer display of gears and cams cannot be taken for
granted because of their apparent simplicity. It was shown in this
chapter (and in Chapter 6) that one shaft could cause another to turn by
means of two bodies in pure rolling contact. Cams were then discussed
by this principle. Types of gear displays were then presented, followed
by an explanation of gear terminology necessary to develop computer
displays.

The fundamental display relationships for gear teeth were then
presented and illustrated in Figure 7.1. Other gear elements and display techniques were then discussed and demonstrated through figures
and subroutines. CALL GTOOTH vs. CALL RTOOTH was then discussed based on gear and rack applications. The subroutines necessary
for complete gear design were listed in Table 7.2. Selecting from this
list, a flow diagram for program steps was determined.

The gear study and analysis by computer program was discussed
next by the presentation of:

1. Involute spur gears
2. Cycloidal gears
3. Hypoid-type gear displays

The remaining portion of the chapter was devoted to the presentation of
cam designs. The classical approach to the design of cams included the
many variations of the plate, cylindrical, and miscellaneous—this was
used also in the computer-aided method. The amount of information
presented was, however, limited to the FORTRAN methods needed.

The chapter concluded with several subprograms for the design of
these basic cams and followers. The reader should be aware that the
software presented is limited to the IBM 370 model 3033 and CALCOMP
or 4010 DVST.

BIBLIOGRAPHY

Anderson, R. H., Programmable automation: the future of computers
    in manufacturing, University of Southern California, Report No.
    ISI/RR-73-2, 1973.
Brown, N. L., Using a computer-aided graphics system to help design
    and draft auto components, Proceedings of the 14th Design Auto-
    mation Conference, New Orleans, La., 1977.
Chasen, S. H., Geometric Principles and Procedures for Computer
    Graphic Applications, Prentice-Hall, Englewood Cliffs, N.J., 1978.
Cheek, T. B., Improving the performance of DVST display systems,
    Proceedings of the Society for Information Display, SID Inter-
    national Symposium Digest Technical Papers, April 1975.

Evans, J. M., Strategies for modular CAD/CAM systems, Proceedings
of the 15th Numerical Control Society, Chicago, 1978.
Laurer, D. J., CAD/CAM interactive graphics systems designed by
users, Proceedings of the 15th NCS, Chicago, 1978.
Levinson, I. J., Machine Design, Reston Publishing, Reston, Va., 1978.
Parr, R. E., Principles of Mechanical Design. McGraw-Hill,
New York, 1970.
Patton, W. J., Kinematics. Reston Publishing, Reston, Va., 1979.
Paul, B., Kinematics and Dynamics of Planar Machinery. Prentice-Hall,
Englewood Cliffs, N.J., 1979.
Phelan, R. M., Fundamentals of Mechanical Design. McGraw-Hill,
New York, 1970.
Ramous, A. J., Applied Kinematics. Prentice-Hall, Englewood Cliffs,
N.J., 1972.
Ryan, D. L., Computer-Aided Graphics and Design. Marcel Dekker,
New York, 1979.
Spotts, M. F., Design of Machine Elements. Prentice-Hall, Englewood
Cliffs, N.J., 1978.

8

Computer-Aided Design
of Flexible Connectors

When the distance between the driving shaft of a computer display and the driven shaft is too great to be connected by gears or cams, a flexible connector is displayed. The most common replacement for a cam or gear is a pulley. If a pulley is turning at an angular speed about an axis, its outer lipped surface will have a linear speed which can be computed in the display program. Its linear speed will result from the input of angular pulley speed and the diameter of the pulley.

COMPUTER SOLUTION OF PITCH SURFACES

If a pair of pulleys are displayed and a flexible connector is stretched between them so that there is enough friction between the connector and the surfaces of the pulleys to prevent slipping; the connector will be displayed in motion with a linear speed approximately equal to the surface speed of the driver. This then will impart the same linear speed to the surface of the follower. The display program must provide animation or turning for the pulleys. The connector may be displayed as a static section, but the observer will assume that the connector is also moving. The display program must also provide for pulleys of multiple sizes on axes that are parallel, intersecting, or neither parallel nor intersecting.

The CAMD user may select flexible connectors for pulleys from several classes.

1. V-belt drives. Here power is transmitted and displayed by the amount of friction present. The amount of power transmitted and displayed on the output device is dependent upon the coefficient of friction between the belt and the pulley. The coefficient of friction assigned by the display program depends

on the materials displayed in contact, their condition (smooth, dry, rough, oily), the arc of contact between the belt and the pulley or sheave, and the display velocity selected for the flexible operation. V-belts forms include double-angle, fractional-horse-power, link, standard, and high-capacity.

2. Flat belts. To make sheaves efficient, several different sizes of belts are available for display. The display size of a flat belt is determined by its cross-sectional area. Standard-duty flat belts with large cross sections are used for heavy-duty or high-horsepower displays; small-cross-section belts are used for light duty. There are many cases where two or more small belts can be displayed to provide the proper amount of belt cross section more economically than can one large one. However, if small flat belts are used on a heavy drive display, a large number would probably be required, because of the low horsepower rating per belt. In this case, one or two large flat belts would be more economical than would many smaller ones.

3. Timing belts. These special belts have found increasing use in industrial plants. Some of these applications include drives that require specific timing of related moving parts, whereas others require a positive transfer of power. To display this, timing belt drives are programmed on a tooth-grip principle in much the same manner as are the teeth on a gear. The molded teeth of the belt are designed to make positive engagement with the grooves on a pulley and to enter and leave the grooves in a smooth, rolling manner. Unlike most other types of belts displayed as a flexible connector, timing connectors do not derive their strength from their driving force (friction contact) or thickness.

Speed Ratios for Belt Connections

Whether a belt drive is displayed as a V, flat, or timing type, the following common names and definitions apply:

Driver sheave. Usually the smallest sheave with the highest rate of animation, the driver sheave is displayed on the motor or other drive component supplying the power.

Driven sheave. Usually, the largest sheave with the lowest rate of animation, it is displayed on the machine being driven.

Idler sheave. Displayed to increase the drive tension, detour the belt strand, reduce belt vibration, and provide takeup.

Belt pitch length. The length of the belt at the neutral axis of the display. The neutral axis is located approximately two-thirds of the distance from the bottom (narrow) to the top (wide) portion of the belt.

Sheave pitch diameter. The sheave diameter at a point where the
neutral axis of the belt contacts the sheave. This is where
the belt and sheave speeds are the same. When calculating
V-belt drive sheaves in a display program, the pitch diameter
is always less than the outside diameter of a sheave.

Arc of contact. The number of degrees of wrap or contact of the belt
around the sheave on a display. A reduction in the arc of contact of the belt displayed changes the power-transmitting
capacity of the flexible connector.

Center distance. Displayed as the distance between the centers of
driver and driven shafts. V-belts will work satisfactorily
when displayed on short or long centers. However, to be
most effective, the center distance should be approximately
equal to or slightly less than the sum of the diameters of the
displayed sheaves.

Speed ratio. Usually calculated in a display program by dividing the
turning rate of the high-speed shaft by the turning rate of the
low-speed shaft. In addition, the ratio can be displayed by
dividing the pitch diameter of the larger sheave by that of the
smaller sheave.

Normally, flexible connectors are displayed from one of three
separate groups, identifiable by the size and shape of the belts. In the
standard display group, belt sizes are currently designated by the
letters A, B, C, D, and E, as shown in Figure 8.1. Each belt has

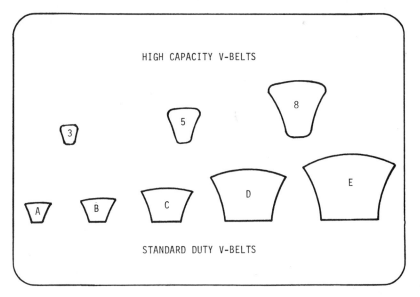

Figure 8.1   Standard-duty and high-capacity V-belts.

specific size limitations, indicated by the full-scale dimensions of the output in Figure 8.1. Some belt manufacturers' sizes vary slightly from the computer model shown. Belts are one type of flexible connector that is manufactured in specific lengths, although they are occasionally purchased as a single strand and spliced to the desired length. The standard belts displayed in Figure 8.1 are still the most common used in machine design.

The second group of flexible connectors that are displayed as computer output can be classified as high-capacity belts. These are used where standard belts may not perform well because of high horsepower or loading conditions. Also, heat, moisture, or other conditions may require a different display. Sometimes, there is no room for a standard drive, and the reduced section of the high-capacity belts allows it to fit into the smaller space. In comparing the outputs shown in Figures 8.1 and 8.2, the user will note that the standard belts are considerably wider than they are high, whereas the high-capacity belts are almost as high as they are wide. Because of the difference in belts and how they must be displayed as computer output, separate display sheaves are used for the standard and high-capacity types.

In addition to the standard and high-capacity belt displays, there are a number of smaller belts used for lighter duty and with smaller drive pulleys. The two types commonly displayed are the L and M series. The 2L through 5L belts are similar in cross section to standard belts and are displayed as factored standard flexible connectors. The 3, 5, 7, and 11M belts are made with a different belt configuration, allowing them to flex more easily and are displayed as factored high-capacity connector belts.

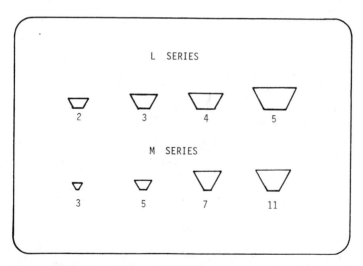

Figure 8.2   L and M series belts.

The value of these types of flexible connectors is important when the designer considers the application.

For example, a running portion of a belt is assumed to have no irregularities in the makeup of the belt, and the upper surface is parallel to and equal in length to the inside surface. When this same section of belt is stretched around a pulley, the inside surface is drawn firmly against the surface of the pulley while the outside surface bends over a circle whose radius is greater than that of the surface of the pulley by an amount equal to the belt thickness. The outer part of the belt must therefore stretch somewhat and the inner part compress. There will also be a section between the inner and outer surfaces that is neither stretched nor compressed. This section may be labeled "NEUTRAL" on the computer display. In a flat belt the neutral section may be assumed to be halfway between the outer and inner surfaces. An imaginary cylindrical surface around the pulley, in which the neutral section of the belt is displayed, is labeled the "PITCH SURFACE" of the pulley. The radius of the pitch surface is used in display programs to represent the effective radius of the display pulley. Therefore, a line in the neutral section of the display belt, at the center of its width, is the line of connection between two pulleys and is tangent to the pitch surfaces; it coincides with a line in each pitch surface known as the pitch line.

In a display program to output Figure 8.3, let the diameter of pulley A be the storage location DIAA, the diameter of pulley B be DIAB, and the half-thickness of the display belt be storage location P. Now assign the speed in rpm of pulley A to storage location VA, and assign the speed of pulley B to VB.

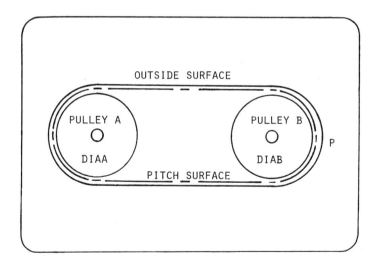

Figure 8.3   Speed ratio and pitch surface.

Then from the display programming in Chapter 2,

linear speed of pitch surface of $A = \pi * VA(DIAA + 2*P)$

and

linear speed of pitch surface of $B = \pi * VB(DIAB + 2*P)$

The belt speed is supposed to be equal to the speed of the pitch surfaces of the pulleys:

$$\pi * VA * (DIAA + 2*P) = \pi * VB * (DIAB + 2*P)$$

or

$$VB/VA = (DIAA + 2*P)$$

That is, the angular speeds of the shafts are the inverse ratio of the effective diameters of the pulleys, and this ratio is constant for circular sheaves. Since the thickness of L and M belts generally is small as compared with the diameters of the sheaves, it may be neglected on the display.

The speed ratio will then become

$$VB/VA = DIAA/DIAB$$

which is the relationship almost always used in the display program calculations.

Pulley-and-Belt Combinations

While there are two major and two minor types of flexible connectors commonly displayed, there are three types of sheaves displayed with them. The sheaves for standard and high-capacity belts are, of course, two of the three display types. The third type is a combination sheave used with both types A and B of the standard belts. Combination sheaves are frequently used in computer-aided machine design having both A- and B-type drives with similar ratios. This allows interchangeability between drives and reduces the number of spare parts in the computer inventory and selection program.

It should be noted in Figure 8.4 that the outside diameter of the sheave is never the same as the pitch diameter. This is especially true when displaying combination sheaves, because the pitch diameter is located at two different points, depending on which belt is being displayed. In high-capacity sheaves, the belt is made so that it extends slightly beyond the outside diameter of the sheave. In instances where grouped belts are displayed with standard sheaves, the belt covers the sheave outside diameter except on the two outside edges.

Variable-speed sheaves are displayed in two general types: manual-adjust and spring-loaded types. The manual-adjust sheave has a smaller range of adjustment than the spring-loaded-type. Although they may differ slightly when displayed, the basic design and animation principles are assumed to be the same. Adjustable sheaves are labeled on the CRT dis-

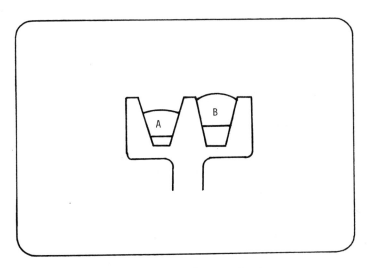

Figure 8.4   Combination A and B standard.

play as manual or spring-loaded. Manually adjusted sheaves are usually displayed on drives requiring only small, occasional adjustment. In these applications, the desired speed is calculated in the display program on the basis of an exact sheave pitch diameter. Then when an adjustable driver sheave is required on the display, small speed adjustments can be made to the driver display to attain a specific speed after the display program is written.

When both the belt profile and sheave design are known, the display program segment for the computation and display of belt and pulleys can be approached. In a display program to output Figure 8.5, let DIAA and DIAB be the storage locations for the connected pulleys, C the storage location for the distance between their axes; and L the length of the belt. Angle Θ is stored in radians. Then the program segment can be written as

$$L=2*(P(1)+P(2)+P(3)+P(4)$$

or

$$L=(\pi/2+\Theta)*DIAA+2*COS(\Theta)+(\pi/2-\Theta)*DIAB$$

or

$$L=\pi/2*(DIAA+DIAB)+\Theta*(DIAA-DIAB)+2*C*COS(\Theta)$$

Of course, a programming language such as FORTRAN will not accept symbols like Θ or π, so the programmer must use THETA and PI as storage locations. And the length of the belt may be stored in a location called BELTL to avoid a mixed-mode expression. Pulley-and-belt combinations are not limited to the output example shown in Figure 8.5. An interactive display program should provide the user with several pulley-and-belt combinations:

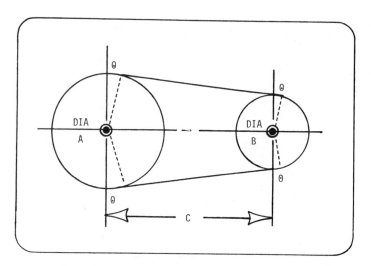

Figure 8.5   Belt-length display.

1. Single pulley, parallel axes
2. Stepped pulleys, parallel axes
3. Open and crossed belt combinations
4. Speed cones and guide pulleys
5. Nonparallel cases for all of the above

The programmer of such a display module provides for the connection of nonparallel shafts by a flat belt. The pulleys must be located so as to conform to a display model for pulleys and belts. That is, the point where the pitch line of the belt leaves a pulley must lie in a plane passing through the center of the pulley which the belt runs. For display purposes, a belt leaving a pulley may be drawn out of the plane of the pulley, but when approaching a pulley its center line must lie in the midplane of that pulley.

COMPUTER DISPLAY OF DRIVERS AND MACHINE ELEMENTS

Computer-generated displays for flexible connections have special tasks to perform in addition to the previously stated functions for the V-belt and a portion of a single-grooved pulley, called the sheave. It is true that these types of flexible connectors are ideally suited for transmitting power from one mechanism to another while compensating for misalignment and allowing for end float in a shaft. When flexible connectors are considered in a larger sense—chains and sprockets, cables and drums, or motors and couplings—a connector must accomplish several objectives while performing its work:

1. Dampening vibration
2. Dampening or absorbing torque
3. Insulating the connection havles from electrical current transfer

Most connectors combine several of these basic purposes in their design. For example, some flexible connectors will do all of the above, whereas others may do only the basic tasks (transmitting power, compensating for misalignment, and end float) and will not insulate or dampen torque or vibration. Still other connectors will not accept misalignment between the shafts. Usually, a CAMD user will select a connector manufacturer that will make several variations of a basic connector that can be displayed for special applications.

Variable-Speed Transmission

The stepped pulleys and speed cones previously listed may be regarded as the elementary mechanisms through which one shaft displayed at a constant speed may drive another shaft at a variety of speeds. In both these cases the speed is changed by program translation to move the belt along the axes of the pulleys to make use of a different pair of working diameters on the driving display and driven display.

Several devices accomplish a similar purpose in a different way. One of these is the variable-speed reducer. Variable-speed reducers perform many functions in the computer-aided design of industrial equipment. A few applications include the control of various speeds in machines or mechanisms, such as turning, blending, and mixing. In addition, they are used in the design of several different mechanisms which are placed in line to assure a smooth, equal speed for the power train. In a simple case, the speed adjustment is accomplished by a variable-speed pulley on the output shaft of a motor.

The selection of a specific type of variable-speed drive to be displayed on a computer output device is governed by many factors. The drive may be displayed by the amount of variable speed required, the speed range, the drive horsepower, or the operating conditions (moist, chemical, dust). Selection is often based on the size of the unit, weight, operating life, reliability of operation, speed response, or cost. Belt-driven variable-speed drives are designed by several manufacturers. The display differences of the various units are minor, usually involving only the display of different bearings or adjusting mechanisms.

The display of the unit is relatively simple. The input shaft of the drive is directly connected to the motor rotor and is supported on either side of the drive pulley by bearings. Usually, the display area of the input shaft that the drive pulley contacts will be displayed as splined or keyed. The spline assures that positive power is transmitted between the shaft and the adjustable sheave.

The input pulley flange width is displayed as adjustable and controls the output speed of the drive by causing the belt to change its position on the pulley flanges. With the belt at the lower position on the flanges, as shown in Figure 8.6A, the reduction ratio would be greatest when the output shaft is rotating at its slowest speed. As the flanges of the pulley are brought closer together, the belt changes its position to the outer edges of the flanges. This is accomplished by the pressure against the belt. The change in position of the belt increases the output speed of the drive shaft. Like the mechanically adjustable V-belts discussed earlier, the driven pulley of the drive is spring-loaded. This helps maintain constant pressure on the belt as it contacts the driven pulley. At the same time as the belt contacting the input pulley is forced to the other extremities of the flanges, as shown in Figure 8.6B, the belt on the driven pulley must position itself. This is accomplished by the use of a spring-loaded driven pulley. As the belt moves closer to the center of the shaft, the springs allow the flanges to move farther apart, giving the belt the needed room and allowing it to move inward.

In addition to the mechanical-type control, some variable-speed units are displayed with pneumatic or electric control. The pneumatic-type display consists of an air cylinder mounted on the drive unit housing. The piston rod end is connected to the movable-flange shifting mechanism. As air pressure is applied to the cylinder, the piston movement causes the shifting mechanism to move the adjustable flange. The electric control unit heads are usually displayed as small reversible motors to cause

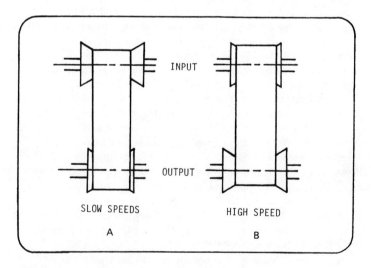

Figure 8.6   Variable-speed drive.

the rotation of the adjusting mechanism normally accomplished by the handwheel in the manual display units.

Most drive units are displayed with openings for ventilation. These are required because the motor and the drive belt must run in a dry atmosphere. Condensation or moisture that generally builds up in a drive case would be harmful to the belt operation. The bearings supporting the shafts are usually displayed as the ball-bearing type. These might be single- or double-row displays, depending on the design of the drive. On occasion, some CAMD users specify tapered roller bearings in some portions of the drive in addition to the ball bearings. However, the forces set up by the driving belt do not include much thrust loading and therefore do not require thrust-absorbing bearings.

Because variable-speed pulleys have at least one movable flange, some means must be displayed to assure that the flange will not freeze to the shafts. Even slight adjustments to the flanges and the drive speed should occur easily. To accomplish this, some CAMD designers allow small amounts of grease to be used on the shafts, but the majority display oil-less, prelubricated, or graphite bushings and/or plastic bearings.

Even though most belt-driven variable-speed units can be designed to accomplish a large speed reduction ratio, they frequently are furnished with additional fixed-speed gear reduction heads. These are designed to bring the reducer into a specific operating range without doing it completely through the belt drive.

Chains and Sprockets

Chain drives, unlike V-belt drives, are not displayed with friction to aid in transmitting motion. Their means of transmitting motion is positive and similar to a gear tooth contact. Because of the positive transfer of motion, chain drives are displayed as totally efficient (98% actual). The chain serves as a flexible connection between the driver and the driven sprockets, allowing them to be spaced some distance apart. The chain weight forms its own takeup on the loose or slack side of the drive. This eliminates the adjustment that is required in a V-belt drive to maintain the proper friction contact. But chain drives do stretch, and occasionally the takeup has to be adjusted or a link or two removed from the chain. Another important feature for CAMD is that a chain drive can be positioned in any part of the driving machinery with only minor programming or display problems. This is accomplished by the link design, which allows the user to place the strand of chain in position and then display it together.

Like V-belt drives, chain drives have specific terms describing their various display components. Some of these are quite similar to those for V-belt drives, whereas others are considerably different. A few of the more common terms follow.

Driver sprocket. Usually, the smaller of the two sprockets and the
one having the highest rpm.

Driven sprocket. Usually, the larger of the two sprockets and the
one having the slower rpm.

Chain pitch. The distance (in display units) from the center of one
connecting pin to the center of the next. In chains having a
solid block link, the chain pitch is on alternate spacing.

Center distance. The number of display units between the centers
of driver and driven shafts.

Chain length. The distance from the center line of the connecting
pin at one end of the strand to the empty connecting hole at
the opposite end. Chains can be specified in pitches or display units.

Chain rating. The load in pounds that the chain will satisfactorily
handle over extended periods of time; also called the
recommended working load. Most CAMD users rate their
chains in maximum or average working load.

Ultimate strength. The strength of the chain before it will break.
This is not a governing factor in the selection of the chain.
However, it gives the shock capacity of the chain.

Pitch diameter. A theoretical circle described by the center link
of the chain as it passes over the sprocket. The pitch diameter
of a sprocket is usually below the top of the tooth or the outside diameter of the sprocket. On drives displaying shortened
teeth, such as silent chains, the pitch diameter may extend
beyond the top of the tooth form.

Because of their wide usage throughout machine design, standards
have been set up for roller chains. Table 8.1 shows the ANSI (American
National Standards Institute) standard roller chain number and the corresponding pitch length for each number. Because of this standardiza-
tion, chains may be designed and manufactured by one firm and inter-
changed with those bearing a similar identification number of another
company. The roller chain is so named because the rollers that contact
the sprocket teeth revolve around a bushing. This turning action allows
the roller to make a rolling contact rather than a sliding contact with the
sprocket teeth, thus decreasing chain wear. Internal sliding action takes
place between the roller and the bushing.

Roller chains are made up of simple display geometry (plate construction) connected by pins with rollers between the side plates. The
side plates are displayed as either the pin link or the roller link. The
pin links are located outside the roller links and connect the roller links
together. Because of this alternate pin/roller link combination, the
chain is normally inventoried (computer part location) in an even number
of pitches. If even pitches cannot be used, an offset connecting link,
sometimes called a half-link, can be displayed to make up one pitch. As

Table 8.1   Standard Roller Chains

Number	Chain pitch (display units)	Chain pitch (mm)	Chain pitch (in.)
25	32	6	0.250
35	48	9	0.375
40	64	12	0.500
50	80	16	0.625
60	98	19	0.750
80	130	25	1.000
100	162	32	1.250
120	195	38	1.500
140	227	44	1.750
160	260	51	2.000
180	292	57	2.250
200	325	63	2.500

shown in the computer display of chain parts (Figure 8.7), the pins extend
all the way through the pin links. Usually, they are riveted to one side
plate, passed through the roller links, and then riveted or connected by
other means, such as cotter pins or spring clips. The roller link is
formed by slipping the rollers on the bushings and the pressing the roller
links onto the bushings. This procedure displays the roller link as a
solid part.

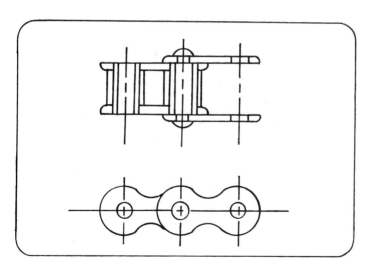

Figure 8.7   Roller chain display.

Silent roller chains have a lace pattern when displayed. The
driving side of the chain forms a tooth contour. This tooth contour
contacts both the leading and trailing edges of the sprocket teeth during
operation. Silent chain drives are displayed in the design of driving
pumps, fans, blowers, and other heavy machinery. Because of their
positive tooth engagement on the sprocket, they are frequently displayed
for timing chain drives.

Silent chains are available from 4-mm pitch to 51-mm pitch,
ranging in width from approximately 6 mm up to 500 mm. The connecting pin design and installation methods vary with different designers.
Silent chains, as with roller chains, have a master or connecting link
that facilitates each installation and removal of the chain for computer
output. In the display of chain transmissions of any kind, horizontal
drives are those having driving and driven shafts in a horizontal plane.
These are always preferable to vertical displays, which have a vertical
center line intersecting the driving and driven shafts. If one sprocket
must be higher than the other, avoid a vertical drive display if possible
by so translating the two sprockets that the common center line inclines
from the vertical as far as is permitted by other conditions that may
govern the design and application. If practicable, an adjustment should
be displayed for the center distance between the driving and driven shafts.
Driving motors are often displayed on adjustable base or slide rails to
provide this adjustment for the center distance.

As a general display rule, the slack strand of a chain should be on
the lower side of a horizontal drive. If the drive is not horizontal but
angular or at some angle less than 90° from vertical, the slack should
preferably be on that side of the display which causes the strand to
curve outward or away from the center line of the driving and driven
shafts. Whenever the slack strand is displayed on the upper side of
either a horizontal or inclined drive, adjustment for the center distance
is especially important, to compensate for possible chain elongation.

Chain-drive sprockets will be displayed as solid body or arm type,
as shown in Chapter 7 for gear displays. Large fabricated steel sprockets
are usually designed with lightening areas to reduce the amount of weight
of the sprocket. Although these are not truly arm-type sprockets, they
are displayed in the same manner, with different notation. Because roller
drive chains sometimes have restricted spaces for their design location,
the hubs are designed in several different styles. These hub designs
should be stored in computer memory as A, B, and C.

As discussed earlier, chain drives are not as flexible as V-belt
drives. For this reason, a degree of caution must be used during the
display programming stage. Actually, displaying a chain drive is not
a complicated procedure, provided that the following simple steps are
observed:

1. Driver and driven shafts should be output level and in parallel alignment. This alignment can be accomplished by graphical translation.*
2. Sprocket alignment should be checked by means of a COMMON statement in the display program.
3. Display the chain with a slight amount of sag on the slack side and the driving side in tension (the chain sag should be approximately 16 display units for every 650 display units of sprocket centers).

Pulley Blocks and Filament Connectors

Power is often displayed as transmitted by means of filament connectors running over pulleys, called blocks, having grooved surfaces. For large amounts of power, such as in a block and tackle, the filaments are made of nylon, rayon, or similar polyester blends. For long-distance drives or heavy loads, wire filament cables are displayed. There are at least two systems of filament connectors that can be displayed:

1. Closed-loop systems, consisting of single or multiple filament strands running between single or multiple pulley blocks. Each strand is parallel (running side by side) in grooves on the pulleys.
2. Open-loop systems, consisting of one filament strand wound around the driving and driven pulleys several times, and conducted back from the last groove of one pulley to the first groove of the other pulley by means of one or more intermediate pulleys, which also serve the purpose of maintaining a uniform tension throughout the entire filament strand.

The shape and proportions of the grooves displayed on many pulley blocks for filament connectors depend somewhat upon the system selected for display. Figure 8.8 shows a display program for interactive CAMD users.

Small-diameter filament strands are often used to connect non-parallel axes, and very often the directional relation of these axes must vary. The most common examples are found in textile machinery such as spinning frames, where the spindles are driven by cords from a long, cylindrical drum whose axis is at right angles to the axes of the spindles. In such cases, the common perpendicular to the two axes must be contained in the planes of the connected pulleys. Both pulleys may be grooved, or one may be cylindrical.

*Graphical translation and other display techniques are discussed in Ryan (1979).

```
//JEWLF8  JOB  (0923-1-410-00- ,:06,2),'WOLFE BX 81'
//STEP1   EXEC    TEK,PDS='JEWLF.BOX',NAME=ROTATE
//C.SYSIN  DD   *
        WRITE(3,1)
1       FORMAT(///,'WHAT SIZE BLOCK DO YOU WANT? (REAL)')
        READ(1,2) SIZE
2       FORMAT(F7.2)
        WRITE(3,3)
3       FORMAT(//,'HOW MANY ROTATIONS DO YOU WANT? (INTEGER)')
        READ(1,4) IROTA
4       FORMAT(I2)
        WRITE(3,5)
5       FORMAT(//,'WHERE DO YOU WANT IT LOCATED?  ENTER TWO REALS',/,
       +'SEPARATED BY A CARRIAGE RETURN.',/)
        READ(1,6)X
        READ(1,6)Y
6       FORMAT(F3.1)
        WRITE(3,7)
7       FORMAT(//,'DO YOU WANT SCREEN ERASED BETWEEN PLOTS?',/,
       +'ENTER 1.  FOR YES,  2.  FOR NO.')
        READ(1,8) TEST
8       FORMAT(F2.0)
        CALL INITT(240)
        THETA=(360./IROTA)*.0174532925
```

Figure 8.8   Display program.

```
      THETA1=0.
      DO 100 I=1,IROTA
      CALL TPLOT(X,Y,3)
      X1=(COS(THETA1)*SIZE)+X
      Y1=(SIN(THETA1)*SIZE)+Y
      CALL TPLOT(X1,Y1,2)
      X2=X1+(SIN(THETA)*SIZE)
      Y2=Y1+(COS(THETA)*SIZE)
      CALL TPLOT(X2,Y2,2)
      X3=X2-(COS(THETA1)*SIZE)
      Y3=Y2-(SIN(THETA1)*SIZE)
      CALL TPLOT(X3,Y3,2)
      CALL TPLOT(X,Y,2)
      CALL TPLOT(X1,Y1,2)
      X5=X1+(SIZE/2.)
      Y5=Y1+SIZE/2.
      CALL TPLOT(X5,Y5,2)
      X6=X2+SIZE/2.
      Y6=Y2+SIZE/2.
      CALL TPLOT(X6,Y6,2)
      X7=X3+SIZE/2.
      Y7=Y3+SIZE/2.
      CALL TPLOT(X7,Y7,2)
      CALL TPLOT(X3,Y3,2)
```

```
      CALL TPLOT(X2,Y2,2)
      CALL TPLOT(X6,Y6,2)
      X8=X+SIZE/2.
      Y8=Y+SIZE/2.
      CALL TPLOT(X7,Y7,2)
      CALL TPLOT(X8,Y8,2)
      CALL TPLOT(X5,Y5,2)
      CALL TPLOT(X8,Y8,2)
      CALL TPLOT(X,Y,2)
      THETA1=THETA1+THETA
      IF(TEST.EQ.2.)GO TO 41
      PAUSE
      CALL ERASE
41    CONTINUE
100   CONTINUE
      CALL FINITT(0,20)
      STOP
      END
      SUBROUTINE TPLOT(X,Y,IPEN)
      IX=X*130.
      IY=Y*130.
      IF(IPEN.EQ.3) GO TO 1001
      IF(IPEN.EQ.2) GO TO 1002
      IF(IPEN.EQ.-3) GO TO 1003
```

Figure 8.8 (continued)

Drum and Cables

When a stranded filament, such as a cable, does not merely pass over a
pulley but is connected to it at one end and then wound upon it; the pulley
is a drum. A drum for a cable is cylindrical and the cable is wound upon
it in helical coils. Each layer of coils increases the effective radius of
the drum by the amount of the diameter of the cable. Wire cables are
very suitable for the transmission of large power for long distances.
The metal cable rigidness, great weight, and rapid destruction due to
bending make it unfit for small-diameter drums or short distances.
Cable will not support the lateral crushing caused by V-shaped sheaves
or similar pulleys. It is therefore necessary to display cables coming
from drums over pulleys with wide grooves so that the cable rests on
the rounded bottom of the groove. The friction is greatly increased in a
display of this type, but the wear of the cable is greatly reduced.

INVENTORY AND SELECTION BY COMPUTER PROGRAM

The design of a flexible connector is a prime case of a user who wants
to deal with CAMD graphics. The problem of using computers in this
field has been limited by two basic shortcomings in our computer tech-
nology and FORTRAN programming:

1. Graphical display—flexible connectors are the prime case—is a
   language all of its own. Translating flexible-connector graphics
   into computer language by hand is tedious, error prone, time
   consuming, and normally more difficult than creating the
   original drawing.
2. Translation of a concept from the designer's mind into graphics
   is complex. It is a series of intricate steps which eventually
   converge as a designer brings many individual and sometimes
   unrelated ideas into the complete and final design.

This portion of Chapter 8 deals with a practical system to meet these
two basic tenets by providing capability such that:

1. The flexible-connector designer deals directly in graphic
   language rather than in an artificial coding language normally
   associated with computer-based automated devices. The user
   is not a programmer; he or she is a designer, engineer, or
   draftsman. The remainder of the chapter will provide a highly
   automated tool to permit the user to work faster, more
   accurately, and more economically.
2. The user of this technique, called inventory and selection by
   computer program, proceeds with the design of a flexible connector as always. Now, however, he or she works at a CRT
   console, and each time a design step is completed, the system

stores each element for evaluation and use again. The user
now has an automated library of connector elements available to
aid, directly and in real time, with the design process. Experience indicates that this technique will significantly increase
the user's performance.

Library of Flexible-Connector Symbols

The CAMD system for flexible connectors accepts graphic data directly.
The user enters the data with a pencil-like electric pen and tablet, shown
in Figure 8.9. To the user it appears as if the pen is actually writing on
the tube face. Actually, the pen and tablet are giving instructions to the
computer, which, in turn, is creating a data sequence that causes the tube
to respond graphically. The computer immediately produces a precise
picture of what it understood the user to mean. This important point ex-
plains why this system is unique. When a user draws a line, this system
immediately produces a picture of that line on the tube face. The picture
shows the user what the computer did in response to his or her request.
The response is what the computer interpreted the designer to mean, not
just a mere copy of the input. If the user approves of what the computer
presented on the console tube face, he or she continues; if not, any mis-
interpretation is immediately corrected. Thus, the user has a continual
running review of what he or she is doing and is building a library of parts
in real-time communication with a computer. The pen/tablet and real-
time feedback are the two powerful and unique features.

The first step a flexible-connector user does is access the library.
This is called "turning the key" or JCL (job control language) in ordinary
computer processing. Because this system is designed especially for
noncomputer operators, an interactive, helping program is tripped by
pressing a single button or typing a single command on the CRT, such as
CALL BEGIN. Figure 8.10 is an example of an interactive, helping pro-

Figure 8.9   Graphics tablet and electric pencil.

```
          TO GET INTO THE TSO SYSTEM ON A CRT THE FOLLOWING STEPS
SHOULD BE DONE (MESSAGE IN SIDE "  "S ARE THE RESPONSE OF THE
TERMINAL TO THE INPUT.) TYPEIN THE FOLLOWING AND ENTER THEM
(BY STRIKING 'ENTER' KEY):
               LOGON
               "ENTER USERID"
               USERID  (YOUR USERID)
               "ENTER CURRENT PASSWORD"
               PASSWORD  (YOUR PASSWORD)
YOU ARE NOW IN THE TSO SYSTEM WHEN "READY" IS DISPLAYED.
          TO REVIEW A PROGRAM OR TYPE IN A PROGRAM YOU HAVE TO GET
IN THE EDIT MODE BY TYPING IN:
               CE .PROG  (PROG IS THE MEMBER NAME YOU GIVE YOUR
                         PROGRAM)
               "NEW MEMBER"  (IF THIS PROG IS A NEW MEMBER)
TO INPUT LINES OF A PROGRAM, HIT THE 'ENTER' KEY. A LINE NUMBER
IS DISPLAY. TYPE IN YOUR LINE AND ENTER. ANOTHER LINE NUMBER WILL
SHOW UP. TO GET OUT OF THE INPUT MODE, HIT 'ENTER' AGAIN. YOU
CAN STORE THE PROGRAM UNDER THIS MEMBER BY ENTER 'S'. TO REVIEW
, ENTER 'L'.
          TO GET OUT OF THE EDIT MODE AND THE TSO SYSTEM ENTER:
               END
               "READY"
               LOGOFF
```

```
          TO DISPLAY GRAPHICAL DATABASE ON THE 4010 DVST. THE FOLLOWING
STEPS ARE TAKEN:
-WHEN USING THE 4010 FOR CALCOMP PREVIEW. THE FOLLOWING JCL IS USED:
  //ABCD JOB (1234-5-678-00-9),'YOURNAME  BOX'
  //SI EXEC PREVIEW,PDS='USERID.PGMLIB',NAME=MYPROG
  //C.SYSIN DD *
      THE PROGRAM WOULD BE PLACED HERE
  /*
  //
WHERE ABCD IS YOUR USERID, 1234-5-678-00-9 IS YOUR ACCOUNT NO.
YOURNAME IS YOUR NAME AND BOX IS YOUR BOX NUMBER. USERID.PGMLID IS
REPLACED BY THE NAME OF THE LOAD LIBRARY YOU ARE GOING TO ALLOCATE
AND NAME IS THE NAME OF YOUR PROGRAM.
ALLOCATE A LOAD LIBRARY BY TYPING IN:
  ALLOC DA('USERID.PGMLIB') SP(1) CYL DIR(1)
     "READY"
  FREEALL
     "READY"
NOW SCHEDULE YOUR PROGRAM BY GETTING INTO THE MEMBER YOU CREATED
AND TYPE IN:
  SCHEDULE
TO DISPLAY YOUR DATA BASE, TYPE IN THE FOLLOWING :
  ATTR ABC INPUT
     "READY"
```

Figure 8.10   Interactive, helping program.

gram response. When applied to lines, circles, curves, letters, and numbers in this fashion, the flexible-connector designer is using real-time communication with a computer to create and revise drawings in the library. If, in the conventional drafting sequence, a horizontal or vertical line is desired, the designer uses a T-square and triangle or standard drafting machine. With this system, the designer pushes a button which trips a subprogram in computer memory, which in turn tells the computer the designer wishes to use the graphics tablet to draw the line. Figure 8.11 instructs the computer to draw only the specified component (straight, vertical, or slant line), no matter how the pen deviates as it is moved across the tablet. Similarly, the user can instruct the computer to draw curves, circles, and arcs.

By depressing the L key from the CRT keyboard and 2 or 3 for dark or bright vector line segment, the designer can position the line with the tablet and pen. Graphical representations of flexible connectors can be built quickly; for example, an entire rectangle can be placed on the face of the CRT with the same effort as a single line simply by depressing the R key and tripping the subprogram listed in Figure 8.12.

The user has a choice of two modes in which to operate:

1. The designer can use the tablet and pen like a pencil and paper to create connector drawings.

2. The designer can use the pen and tablet like a pointer, to activate subprograms that quantitatively define end points of lines, centers of circles, and other boundary conditions at any scale desired. These types of subprograms are presented as menu items (grid of small pictures), appropriately labeled, on a portion of the tablet not used for drawing. The user simply touches the pen to the appropriate grid square location on the tablet to initiate a variety of graphic responses that are cataloged in the user's library. A typical list of library menu items for flexible connector is presented in Table 8.2. In the last column is noted the figure number in this chapter that shows the subroutine for the display item listed.

Since graphic representations are immediately available in mathematical digital form, these data can be used by several different types of peripheral equipment. Such equipment includes microfilm devices, storage (disk) devices, printer/plotters, pen plotters, graphics terminals, and remote viewers. The capability needed and the time and money necessary to create useful software has historically been a pivotal economic consideration for connector users. The programming provided by graphic tablet manufacturers as part of a user system, in conjunction with existing CAMD application programs, will have a considerable impact on this consideration.

```
C*************** PLOT **************************
      SUBROUTINE PLOT(XPAGE,YPAGE,IPEN)
      IX=XPAGE*130
      IY=YPAGE*130
      IF(IPEN.EQ.3) GO TO 290
      IF(IPEN.EQ.2) GO TO 50
      IF(IPEN.EQ.-3) GO TO 60
      IF(IPEN.EQ.-2) GO TO 70
  290 CALL MOVABS(IX,IY)
      GO TO 80
   50 CALL DRWABS(IX,IY)
      GO TO 80
   60 CALL MOVREL(IX,IY)
      GO TO 80
   70 CALL DRWREL(IX,IY)
   80 RETURN
      END
```

Figure 8.11   Pushbutton L software.

```
C************** RECT **************************
      SUBROUTINE RECT(X,Y,WD,HT,THETA)
      THETA=(3.14159/180.)*THETA
      A=X+COS(THETA)*WD
      B=Y+SIN(THETA)*WD
      C=A-SIN(THETA)*HT
      D=B+COS(THETA)*HT
      E=C-COS(THETA)*WD
      F=D-SIN(THETA)*WD
      CALL PLOT(X,Y,3)
      CALL PLOT(A,B,2)
      CALL PLOT(C,D,2)
      CALL PLOT(E,F,2)
      CALL PLOT(X,Y,2)
      RETURN
      END
```

Figure 8.12   Pushbutton R software.

Table 8.2　Flexible-Connector Menu Items

Tablet location		Display item	Display picture	Figure number
X	Y			
0	770	LINE[PLOT]		8.11
0	790	RECT		8.12
0	810	CIRCLE		8.13
10	770	AROHD		8.14
10	790	CNTRL		8.15
10	810	DASH		8.16
20	770	POOCHE[SHADE]		8.17
20	790	DIMLIN[DIMEN]		8.18
20	810	KEYWAY[KEY]		8.19
30	770	PULLEY		8.3,8.4 8.5
30	790	SPROCK		8.7
30	810	BELT		8.4,8.5
40	770	DASHCR		8.20
40	790	CIRARC[CIRCLE]		
40	810	LEADER		8.21
50	770	LETTER	A .... Z	8.22
50	790	NUMBER	0 9	
50	810	SYMBOL	. , - ( ) +	
60	770	USER DEFINED		
:	:	:		
1000	810	USER DEFINED		

```
C******************** CIRCLE ************************
      SUBROUTINE CIRCLE(X,Y,R,SANG,N,THETA)
      X=X-R
      SANG=(3.14/180.)*SANG
      XX=R*(1-COS(SANG))
      YY=R*(SIN(SANG))
      DX=X+XX
      EY=Y+YY
      CALL PLOT(DX,EY,3)
      THETA=(3.14/180.)*THETA
      THETA1=THETA
      DO 260 I=1,N
      FEE=SANG+THETA
      PX=R*(1-COS(FEE))
      SY=R*(SIN(FEE))
      DX=X+PX
      EY=Y+SY
      CALL PLOT(DX,EY,2)
  260 THETA=THETA+THETA1
      Y=0
      X=0
      R=0
      SANG=0
      N=0
      THETA=0
      RETURN
      END
```

Figure 8.13   CIRCLE menu lisiting.

```
C**************** AROHD ************************
      SUBROUTINE AROHD(X,Y,THETA)
      CALL PLOT(X,Y,3)
      A=THETA+180.-15.
      A=(3.14159/180.)*A
      Y1=Y+.125*SIN(A)
      X1=X+.125*COS(A)
      ANG=THETA+180.+15.
      ANG=(3.14159/180.)*ANG
      Y2=Y+.125*SIN(ANG)
      X2=X+.125*COS(ANG)
      CALL PLOT(X1,Y1,2)
      CALL PLOT(X,Y,3)
      CALL PLOT(X2,Y2,2)
      RETURN
      END
```

Figure 8.14   AROHD menu listing.

```
C******************* CNTRL **************************
      SUBROUTINE CNTRL(X,Y,TLEN,THETA,DASH,SPACE,ALINE)
      THETA=3.14159/180.*THETA
      TOTAL=DASH+SPACE+SPACE+ALINE
      NUM=TLEN/2./TOTAL
      CALL PLOT(X,Y,3)
      X1=X
      Y1=Y
      DO 90 I=1,NUM
      X1=X1-DASH/2.*COS(THETA)
      Y1=Y1-DASH/2.*SIN(THETA)
      CALL PLOT(X1,Y1,2)
      X1=X1-SPACE*COS(THETA)
      Y1=Y1-SPACE*SIN(THETA)
      CALL PLOT(X1,Y1,3)
      X1=X1-ALINE*COS(THETA)
      Y1=Y1-ALINE*SIN(THETA)
      CALL PLOT(X1,Y1,2)
      X1=X1-SPACE*COS(THETA)
      Y1=Y1-SPACE*SIN(THETA)
      CALL PLOT(X1,Y1,3)
      X1=X1-DASH/2.*COS(THETA)
      Y1=Y1-DASH/2.*SIN(THETA)
   90 CALL PLOT(X1,Y1,2)
      X1=X
      Y1=Y
      CALL PLOT(X,Y,3)
      DO 100 I=1,NUM
      X1=X1+DASH/2.*COS(THETA)
      Y1=Y1+DASH/2.*SIN(THETA)
      CALL PLOT(X1,Y1,2)
      X1=X1+SPACE*COS(THETA)
      Y1=Y1+SPACE*SIN(THETA)
      CALL PLOT(X1,Y1,3)
      X1=X1+ALINE*COS(THETA)
      Y1=Y1+ALINE*SIN(THETA)
      CALL PLOT(X1,Y1,2)
      X1=X1+SPACE*COS(THETA)
      Y1=Y1+SPACE*SIN(THETA)
      CALL PLOT(X1,Y1,2)
      X1=X1+DASH/2.*COS(THETA)
      Y1=Y1+DASH/2.*SIN(THETA)
  100 CALL PLOT(X1,Y1,2)
      X1=X1+DASH/2.*COS(THETA)
      Y1=Y1+DASH/2.*SIN(THETA)
      RETURN
      END
```

Figure 8.15   CNTRL menu listing.

```
          SUBROUTINE DASH(X,Y,TL,DL,THETA)
          PRINT 1,X,Y
   1      FORMAT(2F6.3,1H| )
   2      FORMAT(2F6.3)
          X1=X
          Y1=Y
          THETA=THETA*0.017453
          M=(TL/DL)/3
          DO 100 I=1,M
          TCOS=DL*COS(THETA)
          TSIN=DL*SIN(THETA)
          X1=TCOS+X1
          Y1=TSIN+Y1
          PRINT 2,X1,Y1
          X1=TCOS+X1
          Y1=TSIN+Y1
          PRINT 1,X1,Y1
          X1=TCOS+X1
          Y1=TSIN+Y1
   100    PRINT 2,X1,Y1
          X1=X
          Y1=Y
          RETURN
          END
```

Figure 8.16   DASH menu listing.

Firmware and Smart Terminals

Firmware provides the instructions to accept data created by the user
(lines, circles, curves, and alphanumerics) and translates these data
automatically at the terminal into digital form. These types of terminals
are called "smart" because the firmware (microprocessor chips) replaces
the need to have subroutines (subprograms) stored in a host computer.
This form is compact and similar to that used in analytical geometry.
For example:

1. A line is stored as coordinates of its end points.
2. A circle is stored as a center point and a radius.
3. A curve is stored as an original point and the three constants
   for a conic equation.

Firmware automatically generates a display code from the pen
location on the tablet and writes this code on a memory disk for use in

```
C**************** SHADE ***********************
      SUBROUTINE SHADE(X,Y,TLEN,THT,S)
      X1=X+TLEN
      Y1=Y+THT
      XX=X1-S
      XS=X1
      YY=Y+S
      YS=Y
      YVAR=Y+S
130   IF(XVAR.LT.X) GO TO 140
      IF(YVAR.GT.Y1) GO TO 150
      CALL PLOT(X1,YVAR,3)
      CALL PLOT(XVAR,Y,2)
      CALL PLOT(X,YY,2)
      XVAR=XVAR-S
      YVAR=YVAR+S
      GC TO 130
140   IF(YVAR.GT.Y1) GO TO 160
      CALL PLOT(X1,YVAR,3)
      YVAR=YVAR+S
      YY=YY+S
      GO TO 140
150   IF(XVAR.LT.X) GO TO 170
      CALL PLOT(XVAR,Y,3)
      CALL PLOT(XX,Y1,2)
      XVAR=XVAR-S
      XX=XX-S
      GO TO 150
160   CALL PLOT(X,YY,3)
      XS=XS-S
      IF(XS.LT.X) GO TO 180
      YY=YY+S
      CALL PLOT(XS,Y1,2)
      GO TO 160
170   CALL PLOT(XX,Y1,2)
      YS=YS+S
      IF(YS.GE.Y1) GO TO 180
      XX=XX-S
      CALL PLOT(X,YS,2)
      GO TO 170
180   RETURN
      END
```

Figure 8.17   SHADE menu listing.

generating the display for designer viewing. These two functions together represent the basic input and output routines and are performed
each time the user changes the pen location. Each element in a connector display is defined as a set of X-Y coordinates imposed on a grid,
with about 1000 X locations and 750 Y locations. As the flexible-connector

```
C******************* DIMEN ************************
      SUBROUTINE DIMEN(X,Y,HT1,HT2,XLINE,THETA)
      CALL PLOT(X,Y,3)
      THETA=3.14159/180.*THETA
      A=X+COS(THETA)*XLINE
      B=Y+SIN(THETA)*XLINE
      CALL PLOT(A,B,2)
      C=A-SIN(THETA)*HT2
      D=B+COS(THETA)*HT2
      CALL PLOT(C,D,3)
      C1=A+SIN(THETA)*.125
      D1=B-COS(THETA)*.125
      CALL PLOT(C1,D1,2)
      E=X-SIN(THETA)*HT1
      F=Y+COS(THETA)*HT1
      CALL PLOT(E,F,3)
      E1=X+SIN(THETA)*.125
      F1=Y-COS(THETA)*.125
      CALL PLOT(E1,F1,2)
      CALL PLOT(A,B,3)
      AR1=A-SIN(1.8326-THETA)*.125
      AR2=B-COS(1.8326-THETA)*.125
      CALL PLOT(AR1,AR2,2)
      CALL PLOT(A,B,3)
      AR3=A-COS(.2618+THETA)*.125
      AR4=B-SIN(.2186+THETA)*.125
      CALL PLOT(AR3,AR4,2)
      CALL PLOT(X,Y,3)
      AR5=X+COS(.2186+THETA)*.125
      AR6=Y+SIN(.2186+THETA)*.125
      CALL PLOT(AR5,AR6,2)
      AR7=X+SIN(1.8326-THETA)*.125
      AR8=Y+COS(1.8326-THETA)*.125
      CALL PLOT(AR7,AR8,3)
      CALL PLOT(X,Y,2)
      RETURN
      END
```

Figure 8.18   DIMEN menu listing.

designer uses the tablet and pen to create elements of a connector display, firmware treats these elements as coordinate-defined points in this grid and adds them to the number of such elements already in computer storage. The designer can ask the terminal to change the size of elements or groups of elements. The terminal, using firmware, expands or contracts

```
C************ KEY *****************************
      SUBROUTINE KEY(XPAG,YPAG,RS,SIZEK,THETA)
      X=(XPAG+RS)-(SIZEK/2.)
      Y=YPAG-SIZEK/2.
      CALL RECT(X,Y,SIZEK,SIZEK,THETA)
      RETURN
      END
```

Figure 8.19   KEY menu listing.

```
C************ LEADER *************************
      SUBROUTINE LEADER(X,Y,X1,Y1,X2,Y2)
      CALL PLOT(X,Y,3)
      CALL PLOT(X1,Y1,2)
      CALL AROHD(X1,Y1,X2,Y2,.125,0.,16)
      RETURN
      END
```

Figure 8.20   DASHCR menu listing.

```
C****************** DASHCR ***********************
      SUBROUTINE DASHCR(X,Y,R)
      X=X-R
      SANG=.1047
      DO 270 I=1,20
      XX=R*(1-COS(SANG))
      YY=R*SIN(SANG)
      DX=X+XX
      EY=Y+YY
      CALL PLOT(DX,EY,3)
      THETA=.1047
      THETA1=THETA
      DO 280 J=1,2
      FEE=SANG+THETA
      PX=R*(1-COS(FEE))
      SY=R*SIN(FEE)
      DX=X+PX
      EY=Y+SY
      CALL PLOT(DX,EY,2)
  280 THETA=THETA+THETA1
      SANG=SANG+THETA
  270 CONTINUE
      RETURN
      END
```

Figure 8.21   LEADER menu listing.

```
C***********************LABEL********************
      SUBROUTINE LABEL(XX,YY,HT,DEG,ICH,NCH)
      DIMENSION ICH(80),IC(44)
      DATA IC/1HA,1HB,1HC,1HD,1HE,1HF,1HG,1HH,
     +        1HI,1HJ,1HK,1HL,1HM,1HN,1HO,1HP,
     +        1HQ,1HR,1HS,1HT,1HU,1HV,1HW,1HX,
     +        1HY,1HZ,1H1,1H2,1H3,1H4,1H5,1H6,
     +        1H7,1H8,1H9,1HO,1H+,1H-,1H ,1H.,
     +        1H,,1H=,1H(,1H)/
      AK=1.
      WD=HT/2.
      THETA=.017453*DEG
      DO 200 II=1,NCH
      AA=II
      DO 230 JJ=1,44
      IF (ICH(II) .NE. IC(JJ))GO TO 230
      IF(II .NE. 1)GO TO 240
      X1=XX
      Y1=YY
      GO TO 250
230   CONTINUE
      STOP
240   X1=XX+COS(THETA)*(((AA-AK)*HT)+((AK-1.)*WD))
      Y1=YY+SIN(THETA)*(((AA-AK)*HT)+((AK-1.)*WD))
250   GO TO (1,2,3,4,5,6,7,8,
     +       9,10,11,12,13,14,15,16,
     +       17,18,19,20,21,22,23,24,
     +       25,26,27,28,29,30,31,32,
     +       34,35,36,37,39,40,41,42,
     +       43,44,45,46),JJ
1     CALL A(X1,Y1,HT,DEG)
      GO TO 200
2     CALL B(X1,Y1,HT,DEG)
      GO TO 200
3     CALL C(X1,Y1,HT,DEG)
      GO TO 200
4     CALL D(X1,Y1,HT,DEG)
      GO TO 200
5     CALL E(X1,Y1,HT,DEG)
      GO TO 200
6     CALL F(X1,Y1,HT,DEG)
      GO TO 200
7     CALL G(X1,Y1,HT,DEG)
      GO TO 200
8     CALL H(X1,Y1,HT,DEG)
      GO TO 200
9     CALL I(X1,Y1,HT,DEG)
      GO TO 200
10    CALL J(X1,Y1,HT,DEG)
      GO TO 200
11    CALL K(X1,Y1,HT,DEG)
      GO TO 200
12    CALL L(X1,Y1,HT,DEG)
      GO TO 200
13    CALL M(X1,Y1,HT,DEG)
      GO TO 200
14    CALL N(X1,Y1,HT,DEG)
      GO TO 200
15    CALL O(X1,Y1,HT,DEG)
```

Figure 8.22　LABEL menu listing.

```
          GC TO 200
16        CALL P(X1,Y1,HT,DEG)
          GC TO 200
17        CALL Q(X1,Y1,HT,DEG)
          GC TO 200
18        CALL R(X1,Y1,HT,DEG)
          GC TO 200
19        CALL S(X1,Y1,HT,DEG)
          GC TO 200
20        CALL T(X1,Y1,HT,DEG)
          GC TO 200
21        CALL U(X1,Y1,HT,DEG)
          GC TO 200
22        CALL V(X1,Y1,HT,DEG)
          GC TO 200
23        CALL W(X1,Y1,HT,DEG)
          GO TO 200
24        CALL X(X1,Y1,HT,DEG)
          GC TO 200
25        CALL Y(X1,Y1,HT,DEG)
          GC TO 200
26        CALL Z(X1,Y1,HT,DEG)
          GO TO 200
27        CALL ONE(X1,Y1,HT,DEG)
          GC TO 200
28        CALL TWO(X1,Y1,HT,DEG)
          GC TO 200
29        CALL THREE(X1,Y1,HT,DEG)
          GO TO 200
30        CALL FOUR(X1,Y1,HT,DEG)
          GC TO 200
31        CALL FIVE(X1,Y1,HT,DEG)
          GO TO 200
32        CALL SIX(X1,Y1,HT,DEG)
          GC TO 200
34        CALL SEVEN(X1,Y1,HT,DEG)
          GC TO 200
35        CALL EIGHT(X1,Y1,HT,DEG)
          GC TO 200
36        CALL NINE(X1,Y1,HT,DEG)
          GC TO 200
37        CALL ZERO(X1,Y1,HT,DEG)
          GO TO 2C0
39        CALL PLUS(X1,Y1,HT,DEG)
          GC TO 200
40        CALL MINUS(X1,Y1,HT,DEG)
          GO TO 200
41        AK=AK+1.
          GC TO 200
42        CALL PERIOD(X1,Y1,HT,DEG)
          GC TO 200
43        CALL COMMA(X1,Y1,HT,DEG)
          GC TO 200
44        CALL EQUAL(X1,Y1,HT,DEG)
          GC TO 200
45        CALL PARENL(X1,Y1,HT,DEG)
          GO TO 200
46        CALL PARENR(X1,Y1,HT,DEG)
200       CCNTINUE
          RETURN
          END
```

Figure 8.22 (continued)

the element by multiplying the lengths of lines, circles, or other connector geometry by a constant factor. The location of elements can be changed through firmware called PAN, which specifies a new location for the set of coordinates describing the element. These simple manipulations are changes handled by the terminal, not the host computer. Firmware performs this bookkeeping inside the terminal while the user initiates the changes and sees the results of his or her action graphically on the console tube face.

The user is not concerned with the inner workings of the terminal, but instructs its action on the console tube face with the pen and tablet. The designer thus avoids program manipulations in awkward languages or the insertion of data via cards, tapes, or typewriters. In fact, the designer need no longer be a computer-oriented person but can act simply as an engineer, tool designer, layout person, or draftsman. The time, and therefore cost, savings resulting from bypassing these procedures are substantial.

SUMMARY

The CAMD user often has need for a technique to display a large number of small specialized parts representing an organized mechanism. The design of flexible-connector systems is just such a case in point. As many as 35 separate display elements may go into the single display of a driver and driven element if flexible connectors are used. The reader is introduced to the design of flexible connectors for pulleys from several classes: V-belt, flat, and timing.

Speed ratios for belt connections were discussed together with the definition of the basic pulley and belt terminology used in CAMD. A group of examples were shown as computer displays. Next, pulley combinations, such as single, stepped, speed cones, and nonparallel cases, were discussed.

Variable-speed transmission examples were presented using the flexible connector to dampen vibration and reduce shaft speeds. How to display and design chains and sprockets were introduced, together with the CAMD terminology. Standard roller chains and how to represent them in a computer display were included, as were pulley blocks and filament connectors. Here an interactive program technique was presented. Drums and cables are the most popular use of stranded filament connectors.

In the final section the reader was introduced to inventory and selection programming for flexible connectors. This was presented in two sections: a library of flexible-connector symbols, and the firmware approach. A flexible connector menu list was also presented and displayed.

BIBLIOGRAPHY

Andrews, H. C., Computer, digital image processing, IEEE, May 1974.
Boguslavsky, B. W., Elementary Computer Programming in FORTRAN.
    Reston Publishing, Reston, Va., 1974.
Cheek, T. B., Improving the performance of DVST display systems,
    Proceedings of the Society for Information Display, International
    Symposium Digest Technical Papers, April 1975.
Croft, F. K., Why use computer graphics? Proceedings of the
    Computer Graphics Seminar, University of Louisville,
    Louisville, Ky., March 1980
Hollingum, J., Computer graphics cuts down drawing board time,
    Engineer, May 1977.
Koenigsberg, L. K., A graphics operating system, Computer Graphics,
    Vol. 9, No. 1, Spring, 1975.
Rembold, U., M. K. Seth, and J. S. Weinstein, Computers in manufac-
    turing. Marcell Dekker, New York, 1977.
Ryan, D. L., Computer-Aided Graphics and Design. Marcel Dekker,
    New York, 1979.

9

Computer-Matched Machine Elements

The use of a digital computer to describe, draw, document, and design a machine element has been discussed in the preceding chapters. It is now the intent to merge all the concepts into a single working system. Overall, this system will provide the programs for creating and revising designs by manipulating lines, curves, and shapes within the rules of plane and solid geometry. Machine elements designed in the X, Y, and Z axis coordinate system are written in modular form so that desirable elements can be matched. To clarify this interplay between an operator, the computer program, and a cathode ray console, a few design situations involving element trains have been chosen.

Those trains chosen represent a working system that allows an operator to make an assumption of a multiple subprogram library and then look at a series of display plots on a scope to evaluate original assumptions. Some 30 to 35 variables must be satisfied to finalize the train and complete the computer match. Design firms using a system like this have found that the time has been reduced from 3 months to 6 or 7 days; and the number of iteratives increased from 3 or 4 to 12 to 15. This 10 or 15:1 time savings has allowed several innovations. First, the design can take place during the proposal stage of a machine project. Second, changes in design are possible with little effect on the schedule of the train selection. Third, errors in prototypes can often be corrected before the final train selection is made.

COMPUTER-SELECTED TRAINS

A train is a series of rolling cylinders or cones, gears, pulleys, or similar devices serving to transmit power from one shaft to another. The reader has been introduced to each of these elements, how to present

each as a single element, and how to design certain "display pairs." A
pair of rolling cylinders, gears, or pulleys is really a train, but usually
the design concept of a train is applied only to those computer-matched
combinations in which there are more than one pair.

This being the case, the designer must do the selection on-line in
real time. Such designers use a large library of preprogrammed
machine elements and are a demanding users group. They expect computer responses to be instantaneous whenever doing a matching task which
they consider trival or obvious (no matter how large the library to be
sorted). They generally become more tolerant when the matching and
combination tasks or requests for responses seem more complex. They
create for the computer, storing the library, a set of intermittent and
somewhat random inputs, and the matching the computer must provide
may be simple or complex, at the momentary decision of the users.

Machine designers are slow by computer standards. Their fastest
response requirements can be supplied by feedback signals with 0.1-sec
time delays. The designer is adaptable and will wait 10 to 30 sec for
feedback on more sophisticated questions. But if the designer must wait
2 to 5 min, their thought processes for matching will wander. The design
firm using computer-aided techniques must try to provide a system that
can interlace and overlap these user requests so that the expensive parts
of the system are busy with a smooth flow of work in spite of the intermittent random nature of the input. This system must take advantage of
the slow demands of a human being in terms of the microsecond speeds
of the computer.

With this condition in mind, two categories of users for computer
matching of machine elements tend to emerge:

1. Those who are working on the same or related problems
2. Those whose applications are independent of others

The first group tend to use the same database and the same set of basic
subroutines so that a computer library holding such routines can be time-
shared by interlacing the requests to run these routines.

The second group is more demanding. Each operator may well ask
for a large amount of computer memory, and each request may demand a
rather extensive change of the contents of memory. This is more like
equipment sharing than time sharing and must be recognized as such. The
first group of train designers can benefit by the rapid communication path
available to several users. The mass of storage that can be justified by
sharing the same information is an asset to all, and the idea of centraliza-
tion and consolidation becomes attractive. Conversely, the CAMD independent user may consider that all this multiple-access time sharing is
an economic necessity that can be tolerated for economic reasons, provided that he or she has sole use of a graphics tablet.

A close look at these demands will reveal that the one basic common denominator running through all applications is the CAMD designer. The machine that responds to the designer (i.e., receives inputs, alphanumeric or graphic, from tablet and electric pencil to produce a display) is not necessarily the same machine that has the capability to solve the problems that the user defines as he or she formulates the design of the train. The advent in the last two or three years of small, fast machines (super-minis) has made it possible to consider coupling users to digital computers in real time at reasonable cost. The large host machine, receiving requests from a group of smaller units, now appears to be more efficiently run.

This idea of small and large machines sharing the load to gain flexibility, speed, and efficiency is foreign to many proponents of large central CAMD systems. The purpose of this chapter is to show that the use of a graphics tablet, coupled with the response time of a system, will be the criterion of the on-line, real-time user and that other yardsticks, such as memory speed, arithmetic speed, size of core, and others, are only the means of gaining this response time. Figure 9.1 represents the outline of the graphics tablet with the console viewing area clearly labeled. Directly above this area is the preprogrammed library of elements from which the train designer is to select. The selection process begins with the designer touching the electric pencil to the elements desired; this trips a preprogrammed library symbol. Next, the user selects an X, Y position inside the DVST console viewing area and presses the electric pen to the tablet. The X and Y location for the preprogrammed library element is now satisfied and the element appears on the face of the CRT. The train designer enters the data with a press of the pen; first selection from the top library and then placement on the viewing area. To the user it appears as if the pen is actually writing on the face of the tube. Operators tend to look at the tablet for library selection and look up at the CRT for placement of the train element.

Driver and Pinion

Figure 9.2 is a representation of a gear train design using the graphics tablet and CRT display methods. Display element 1 is a gear keyed fast to shaft A. Element 2 is a gear keyed fast to shaft B and meshing with 1. Element 3 is another gear also held fast to shaft B and meshing with gear 4, which is keyed to shaft C. To describe the interaction, only those library elements are visible that have been used in the display. The + mark represents the pen location and placement in the viewing area of the CRT. The designer plans to have shaft A turn; 1 will turn with it and cause 2 to turn. Since 2 is keyed to shaft B, B will turn with 2. Gear 3 will then turn at the same angular speed as 2 and will cause 4 to turn, causing shaft C to turn with it. Element 1 is called the driver and 2 the driven or pinion. Similarly, 3 turning with 2 is the driver and 4 the

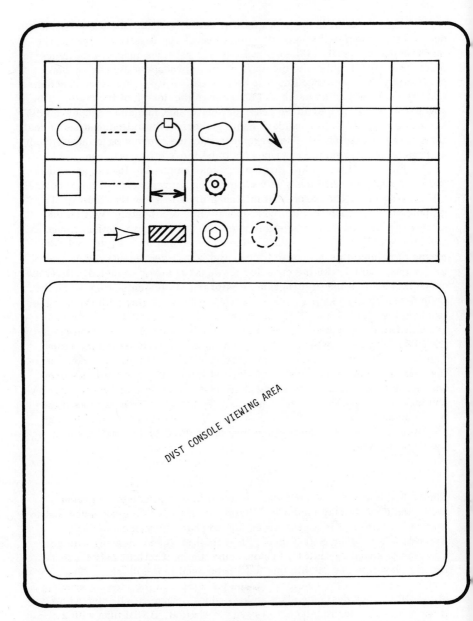

Figure 9.1   Layout of the graphics tablet.

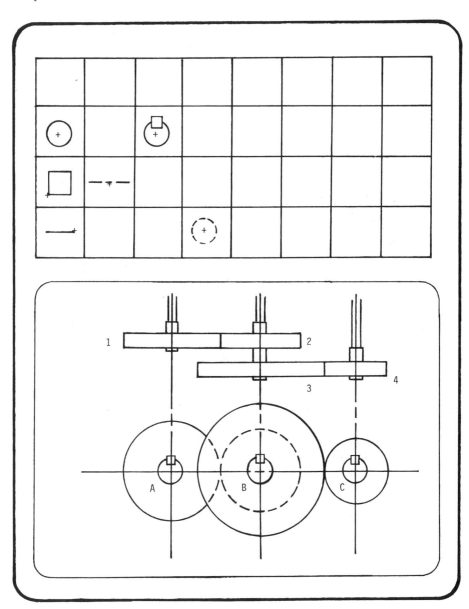

Figure 9.2   Use of tablet for gear trains.

driven. Therefore, in any train displayed like this, consisting of three
axes with two pairs of wheels, two of the wheels are drivers and two are
driven.

Idler and Train Value

Figure 9.3 is the CRT portion of the tablet and pen method of displaying
trains. In this example there are only three gears. Gear 2 drives 3,
which in turn drives 4 along a common axis. Element 3 is therefore
both a driven and a driving gear. Such a gear is called an <u>idler.</u> When
two shafts are connected by two external gears, the shafts will rotate in
opposite directions, but if an idler gear is placed between these two
gears, their direction of rotation will be the same. An idler is also
used to reduce the size of gears required to connect two shafts with a
fixed center distance and a desired velocity ratio. An idler does not
affect the velocity ratio.

  If the fixed piece is considered the member that supports the axes
of the wheels of a train, the <u>train value</u> may be found in the display programs as the ratio of the absolute angular speed of the last wheel or gear
(driven) to the absolute angular speed of the first wheel or gear. The
train value is the reciprocal of the speed ratio as defined in Chapter 7.
The member carrying the wheel axes may be the frame of the machine
or it may be an arm or link which is itself in motion. If the train value
is designated in FORTRAN by TV, then

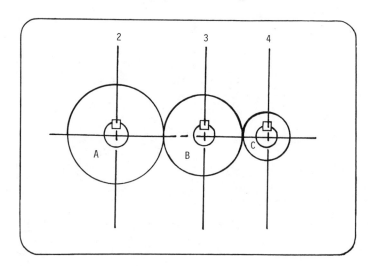

Figure 9.3   Idler display.

TV = 1/SR

where SR is the speed ratio. In Figure 9.1, if all the shafts are in fixed
bearings, shaft A turns at 20 rpm, and the sizes of the several gears are
such that shaft C makes 120 rpm, TV = 120/20 or 6. An inspection of the
same display will indicate that, if A turns clockwise, B will turn counterclockwise and C will turn clockwise. The direction of C is the same as
that of A. The value of this train is said to be positive and will be stored
with a plus sign for TV. If the number of wheels involved is such that the
last shaft turns in the opposite direction from the first shaft, the value of
the train will be stored as negative.

Reverted Gear Trains

When the driving and driven gears have coincidental axes, the gear train
is displayed as a reverted gear train. A typical display would be the back
gear arrangement for a simple cone pulley headstock on an engine lathe.
This illustrates the principles involved when two gears whose axes coincide
are connected by a train of gears through an intermediate shaft, the axis of
the intermediate shaft beign parallel to the axis of the connected gears. In
a FORTRAN display program, storage location P is the cone pulley, which
may run loose on one of the spindles stored under SA (for spindle A). G2
can be a storage location representing an integral gear with P that meshes
with a second gear stored as G3. G4 is another gear location on the same
shaft with G3, both G5 and G4 being keyed to the shaft. G4 meshes with G5
fixed to SA. From Chapter 7:

$$Na = Np * \frac{teeth\ in\ G2\ *\ teeth\ in\ G4}{teeth\ in\ G3\ *\ teeth\ in\ G5}$$

Since the shaft stored as SB is parallel to SA, the gears must be so proportioned that the pitch radius of G2 + pitch radius of G3 equals pitch radius
of G4 + pitch radius of G5. Consequently, if the pitches of the two pairs
are to be in some definite ratio, there must be a corresponding relation
between the sum of the teeth in G2 and G3 and the sum of the teeth in G4
and G5.

Once this has been programmed, the reverted gear train can be output as shown in Figure 9.4. Similar examples can be selected from the
following:

1. Clockworks
2. Fiber carding machines
3. Hosting devices
4. Certain speed drives

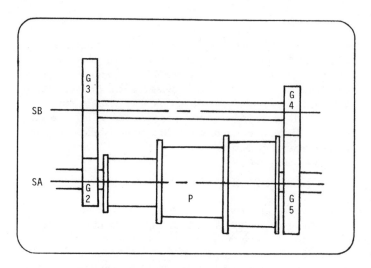

Figure 9.4   Reverted gear train display.

DISPLAY OF EXAMPLE TRAINS

The following three examples—speed gear, automative transmission, and
cone gears—are presented because they serve to illustrate the display
principles discussed earlier, not because a knowledge of these particular
trains is of special computer-aided importance.

Speed Gear Trains

Many drives consist of gears so displayed as to constitute two or more
trains with different train values, any one of which may be used. The
different combinations that may be displayed for this purpose may be
sorted and matched as follows:

1. First and last display elements with axes coinciding. Connections made by sliding gears employing one or more intermediate
   parallel shafts, all shafts in fixed bearings (automotive
   transmission).
2. Driving and driven shafts parallel. One shaft carries several
   gears of different sizes, each keyed to the shaft and displayed
   in mesh with a corresponding gear on the other shaft.
3. Displays made by matching an intermediate parallel shaft whose
   bearings may be quickly adjusted to bring the gears into mesh
   with those on the main axis (reverted gears).
4. Displays on parallel axes, one of which carries several gears
   of different sizes. Matching is made by employing a sliding
   gear on the other axis with an adjustable idler to complete the
   train (machine tools).

Automotive Transmission

Figure 9.5 is a display of a schematic gear train for a manual transmission. The rectangle representing gear A and displayed from storage location GA is the main driving gear. GA receives its motion from the automobile engine through the clutch. GA turns freely on the axis labeled P. Gear D is displayed to turn freely on P. Gear F (GF storage) may be slid along P but is keyed to P so that they turn as a pair. K is the storage location of a clutch which may slide along P but causes P to turn when it turns. GB, GC, GE, and GH are all display gears which turn as a unit on axis S parallel to P. GI is a gear displayed to mesh with GH and turning on an axis in front of and above S. GA is in mesh with GB, and GC with GD, at all times.

   The train is displayed in neutral. If the engine is running and the clutch is in, so that GA is connected to the engine, all the gears except GF are turning; but the axis P with GF and clutch K may be at rest or may be turning independently of the other gears if the car is coasting. Now if the gears are shifted to move GF along P to the left of the display screen, GF will come into mesh with GE. Since GF is keyed to P, the drive is now from GA to GB and from GE, which turns with GB to GF. This action produces low forward gear.

   If the conical hub of K comes in contact with the surface of the space ring, a gradual synchronizing of the speeds of GD and K will occur. This will result in a sliding action which gives the positive drive from GA through GB, GC, and GD to P. This is intermediate or second forward speed.

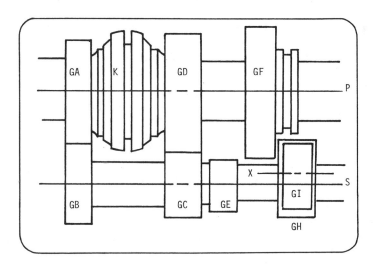

Figure 9.5   Display of a schematic gear train.

Moving K to the left connects K directly with GA, first synchronizing their speeds in the same manner as K and GD. GA is now connected directly to P through K, and P has the same speed as GA. This is the highest forward speed for the gear train. A fourth ratio may be selected by placing K in a neutral position and moving GF to the right of the display. GF comes in mesh with the idler GI, giving the drive from GA through GB, GH, and GI to GF, with the idler GI causing GF and therefore P to turn in the opposite sense from GA. This is the reverse gear.

Cone Gears

These types of trains are often used in machine tools. A good example is the feed train of an engine lathe. The driving shaft of this example receives its motion from the lathe spindle through another train. The driving shaft is provided with a keyway that matches a key on the main driving gear, labeled GA in most computer display programs of this type. The hub of this main driving gear usually fits (turns) in a hole in the adjusting arm with a collar displayed at the side to keep the two together. GA drives an idler which turns on a pin displayed on the arm. This group of display elements may be programmed to move along a position opposite any one of the cone gears in the display. The axis of the idler can be swung about the axis of the spindle by CALL ROTATE until the idler comes into mesh with the cone gears. Since there may be up to 11 gears in the display cone, it is possible to output several different speeds for a given speed of the spindle.

DESIGN OF GEAR TRAINS

No definite rules or formulas for success can be followed in designing a train of gears to have a certain train value. The design process is mainly one of "sort and match" until the desired result is displayed. Certain lines of attack may be followed, however:

1. It may be desirable to have as few gears as possible in order to reduce power losses due to friction in the bearings and between the teeth.
2. Select as many gears alike as possible, to simplify gear cutting operations and reduce the number of sizes to be kept in stock for replacements.
3. Keep the speed change per pair of gears from being excessive. If the train value is selected arbitrarily, it may be necessary to display a greater number of pairs of gears to obtain the exact train value than is visually required to output an approximate train value. In some cases it may be impossible to select gears that will output the exact train value. These methods may best be illustrated by discussing typical examples.

Selected Examples

Suppose that a designer selects the gears for a train whose train value is +16. If no gear is to have fewer than 12 teeth or more than 60 teeth, what is the sorting and matching involved?

The maximum train value per pair is entered to the display as

MAXVAL=MAXNT/MINNT

where MAXNT is the maximum number of teeth (60) and MINNT the minimum number of teeth (12). Therefore, the maximum train value (MAXVAL) is 5. Since 5 is less than 16, one pair is insufficient. If two pairs are used, the maximum possible train value is

$25 = 5*5$

which is greater than 16. The program then selects two pairs for display. If the pairs are to be alike, the display program will compute the square root of the train value, which is 4 in this example.

Now the sorting routine of the program computes (multiplies the numerator and denominator of each pair by the same number so that the product output will not be less than 12 or more than 60) the minimum and maximum number of teeth. There are four possibilities returned from the sorting routine:

$$12 \quad \frac{16}{1} = \frac{48}{12} * \frac{48}{12} \tag{1}$$

$$13 \quad \frac{16}{1} = \frac{52}{13} * \frac{52}{13} \tag{2}$$

$$14 \quad \frac{16}{1} = \frac{56}{14} * \frac{56}{14} \tag{3}$$

$$15 \quad \frac{16}{1} = \frac{60}{15} * \frac{60}{15} \tag{4}$$

for the designer to select.

In another example, a designer must select the gears for a train in which the last gear will turn 23 times while the first gear turns once (train value 23) and the direction of rotation will be opposite (−23). No gear is to have fewer than 12 teeth nor more than 70. The maximum train value per pair is returned from the display program as

$5.8 = 70/12$

which is less than the required train value of −23. Two pairs are now required. If each pair is to be displayed alike, the program selects the train value per pair by computing the square root of 23. The program cannot return an exact square root for 23, so the program next branches to factor

$$\frac{23}{1} = \frac{23}{4} * \frac{4}{1}$$

Finding no way of obtaining the factors, the program returns the square root of 23 (4+). The program uses 4 in the first factor and 23/4 in the second factor, having a resultant of 23/1 after the sorting. Since the factors are unlike, the pairs of gears will be unlike and the program outputs

$$\frac{23}{1} = \frac{69}{12} * \frac{48}{12}$$

by multiplication of both the numerator and denominator of the first factor by 12 and the second factor by 3.

To obtain a negative train value, it will be necessary to display an idler in the train. This idler may be any number of teeth from 12 to 70; the number will depend upon the space requirements.

Matching Problems

Computer-matching techniques involve the mechanisms in a train. A train is a series of rolling cylinders, cones, gears, clutches, and couplings. Whenever a clutch type of mechanism is designed and displayed between driving and driven components in a train, it actually is matched to perform the design of both a clutch and a coupling. While the clutch is disengaged, the two sections of train are independent of each other. This will allow one part to rotate while the other remains stationary. Or in certain applications, the driven part may turn faster or ahead of the driving part.

In the case of fluid or magnetic couplings in trains, the driving and driven components may never come in physical contact. The two components transmit motion through a second medium, which can be either a hydraulic fluid or a magnetic field. When the driven half of these nonmechanical couplings becomes overloaded, it will slip and begin to slow down. At this time, the coupling acts as a clutch.

The matching program for couplings and clutches must select from the following types;

1. Friction clutches, which use a lined metal or fibrous plate mounted between two steel plates for transferring motion between two trains,
2. Jaw clutches, which are usually displayed on slow speed applications. The motion for engaging the clutch is displayed with a shifting arm,
3. Torque-limiting clutches, where the driven half of the clutch is always engaged. The train will usually rotate when it is started, but when the train becomes overloaded the coupling will slip between the two friction plates until the load is reduced,

4. Centrifugal clutches are common in applications where it is desirable to have either no load starting or protection against overloads,

5. Overrunning clutches or one-way clutches are frequently used on trains where the driving motor requires protection.

SUMMARY

It was the intent of this chapter to merge all the concepts of the previous chapters into a single working system. This system provided a method for creating and revising designs by manipulating lines, curves, and shapes within the rules of plane and solid geometry. Machine elements that were designed in X, Y, and Z axis coordinate systems were described in Chapter 8. In this chapter they were presented in modular form so that desirable elements could be matched and displayed. To clarify this interplay among operator, computer program, and cathode ray console, a few design situations involving element trains were presented.

Those examples chosen represented a working system which allowed the designer to make an assumption of a multiple subprogram library and then to look at a series of displayed plots on a scope. Several variables were satisfied to finalize the train design and complete the computer matching.

BIBLIOGRAPHY

Anderson, R. H., Storage cathode-ray tubes and circuits, Tektronix, Inc., Beaverton Ohio, 1968.

Bouguslavsky, B. W., Elementary Computer Programming in FORTRAN. Reston Publishing, Reston, Va., 1974.

Chasen, S. H., Geometric Principles and Procedures for Computer Graphic Applications. Prentice-Hall, Englewood Cliffs, N.J., 1978.

Giloi, W. K., Interactive Computer Graphics. Prentice-Hall, Englewood Cliffs, N.J., 1978.

Newman, W. and R. F. Sproull, Principles of Interactive Computer Graphics, 2nd ed. McGraw-Hill, New York, 1980.

Ryan, D. L., Computer-Aided Graphics and Design. Marcell Dekker, New York, 1979.

10

Computer-Aided Combination of Machine Elements in the Design of Working Machinery .

Each of the preceding chapters has presented a single working element and has shown how to use the digital computer for display, design analysis (loads on machine members), and related items. This final chapter incorporates all of these concepts to help the machine designer decide what device, mechanism, or machine element should be used in the final machine. Figure 10.1 shows a typical setting for this decision.

In an industrial setting, a user finds computer-aided design being employed for new products, apparatus, machines, systems, tools, structures, circuits, and many other applications. In the case of computer-aided kinetics for machine design, this may cover everything from a ball bearing 1 mm in diameter to the machinery for a complete manufacturing plant. It may be used to design and analyze a small piece part or a jet engine for an aircraft. The user of this technique does analytical investigations and tests, and studies various computer programs for modeling real-world situations, as illustrated in Figure 10.2.

The purpose of simulation displays for the dynamic illustration of machine elements as shown in Figure 10.3 is to improve the computer-aided design of manufactured products by American industrial firms. This improvement normally involves those processes that require a description of geometric edges, as displayed in Figure 10.4. The edges are joined together to form surfaces as shown in Figure 10.5, and multi-surfaces are used to describe three-dimensional objects as displayed in Figure 10.6. The computer has been of significant assistance in the display of the design as well as the actual part programming for numerical control (N/C), as shown in Figure 10.7. Here a part programmer simulates the machine tool that will produce the finished part. The part programmer's commands are transferred by the computer-aided machine design (CAMD) software to a suitable N/C readable format, as shown in Figure 10.8. The process described here is not CNC or DNC

Figure 10.1   Mechanical engineer and machine designer discuss final selection of devices, mechanisms, or machine elements for the proposed machine. (Courtesy of Auto-Trol Corp.)

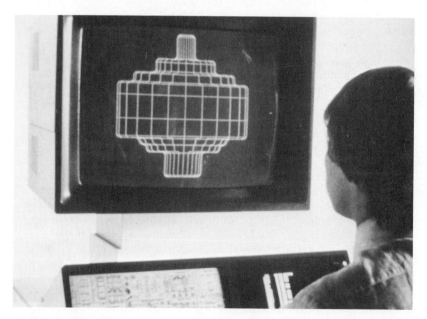

Figure 10.2   User performs analytical investigations and tests, and studies computer program for modeling and simulating real-world mechanical devices. (Courtesy of Auto-Trol Corp.)

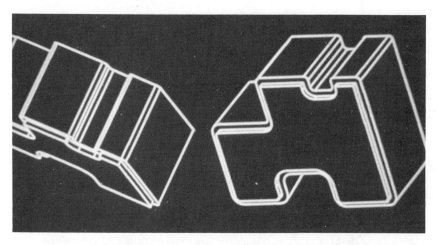

Figure 10.3   Dynamic illustration of part animation on DVST output
device for study of machine part properties. (Courtesy of Auto-Trol Corp.)

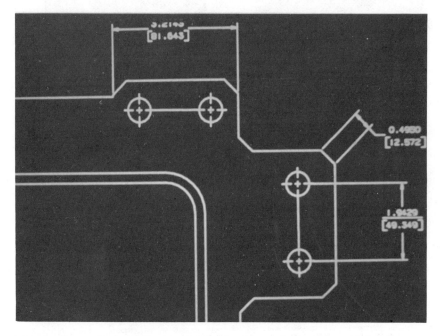

Figure 10.4   Description of geometric edges displayed on DVST.
(Courtesy of Auto-Trol Corp.)

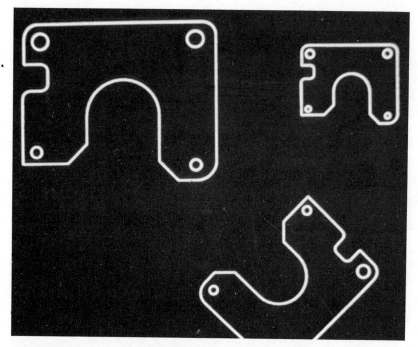

Figure 10.5   Geometric edges joined to form surfaces on DVST.
(Courtesy of Auto-Trol Corp.)

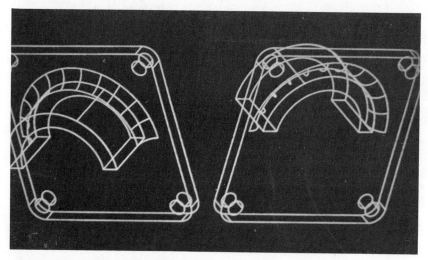

Figure 10.6   Multisurfaces joined to form three-dimensional objects.
(Courtesy of Auto-Trol Corp.)

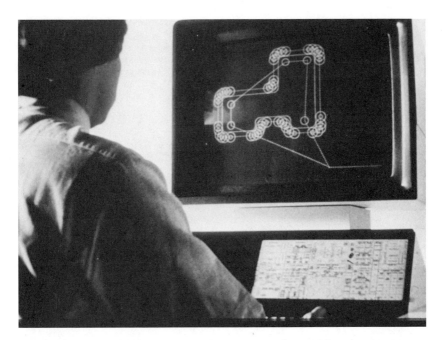

Figure 10.7   Part programming for numerical control (N/C) applications. (Courtesy of Auto-Trol Corp.)

Figure 10.8   Part programmer's commands transfer from CAMD format to N/C readable tape for use on automated machining center. (Courtesy of Auto-Trol Corp.)

but is referred to as CAM (computer-aided manufacturing) processing.
The CAMD system described throughout this text provides test tapes for
the N/C operator shown in Figure 10.9; he loads the machine-readable
format tape produced by the part programmer shown in Figure 10.7.
The output of the tape is then run on the N/C machine pictured in
Figure 10.10.

Automotive companies and the major aerospace firms have developed
computer-aided design systems as just described. They deal with N/C
part animation, simulation, and documentation as well as production.
These systems or programs are very costly and jealously guarded. It
was my contention during the development of programs for the first nine
chapters that there was no reason why all computerized approaches must
be expansive or expensive. Not all design problems are on the scale of
an automobile or airplane, and simple examples were chosen throughout.
If the scale is modest, as in Figure 10.11, an equally modest display can
be developed to handle it.

Figure 10.9   N/C operator loads test tape for CAMD-produced part.
(Courtesy of Auto-Trol Corp.)

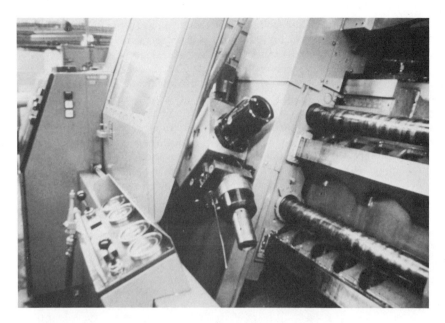

Figure 10.10   N/C machining center. (Courtesy of Auto-Trol Corp.)

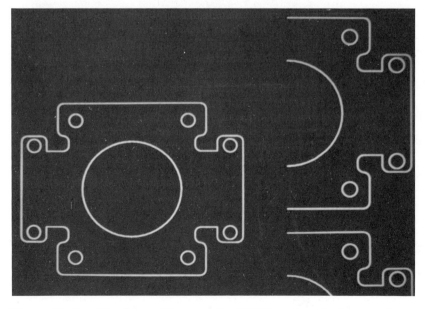

Figure 10.11   Simple machine part for CAMD system design output.
(Courtesy of Auto-Trol Corp.)

These simple approaches described, for example, the technique in
hardware and software used for the display of velocity analysis. The
intermixed display of refresh and direct-view storage tube (618 DUST)
graphics was presented as shown in Figure 10.12, because traditional
storage using cathode ray tubes (CRTs) has the advantage of low cost
coupled with the ability to display large amounts of graphic information.
DVST displays alone do not provide the dynamic motion and transformation
required in the animation and simulation of machine parts; but high-
resolution, refreshed CRTs tend to be very expensive and often require
considerable software. Combining refresh and storage graphics on the
same display allows designers to output static and dynamic areas to
achieve, at reduced cost, outputs such as that shown in Figure 10.13.

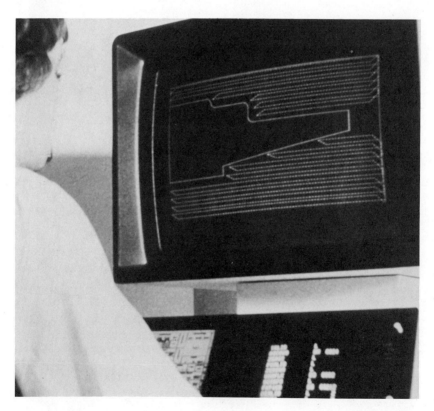

Figure 10.12   Typical CAMD display console. (Courtesy of Auto-Trol
Corp.)

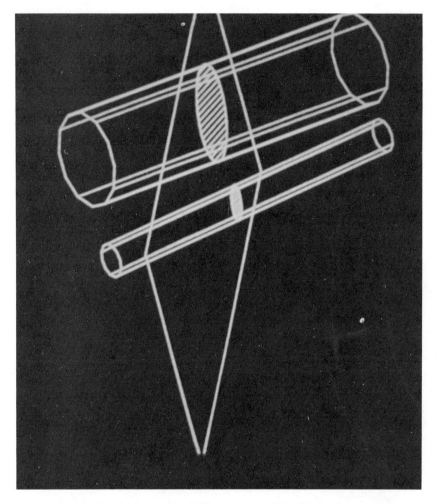

Figure 10.13 Typical graphical display from CAMD display console.
(Courtesy of Auto-Trol Corp.)

With the advent of extremely high speed computers and the associated engineering graphics hardware available today; the machinery
designer can save time and money. The designer is smart, creative,
and slow; whereas the computer and its various graphical displays is
stupid, uncreative, and very fast. The situation is thus a working relationship whereby the person and the machine can interact. Certainly,

their characteristics complement each other, but their languages are
very different. People think in symbols and pictures, whereas the
computer understands only simple electrical impulses. Hardware
innovations such as the graphics tablet; where the designer may enter
a sketch for the computer to store as numerical data based upon a
graphical shape (database), interaction at a design level is now possible.
This interaction can best be demonstrated by a series of figures representing a graphical conversation. Figure 10.14 illustrates a way of translating
human instructions into electronic data, and conversely, the convertion
of the computers impulses into engineering documentation known as a
working drawing, the working drawing resulting from the interaction
that took place. For example, Figure 10.15 is an output of the documentation of the front view only (note that this version is dimensioned
in English, whereas Figure 10.14 used metric). Figure 10.16 is an
output of section AII only (note that this has designer's notes for CAM
included).

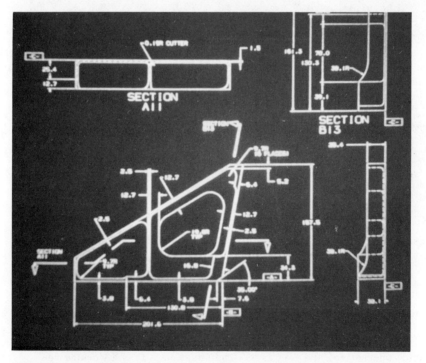

Figure 10.14   Working-drawing output on DVST portion of CAMD
console. (Courtesy of Auto-Trol Corp.)

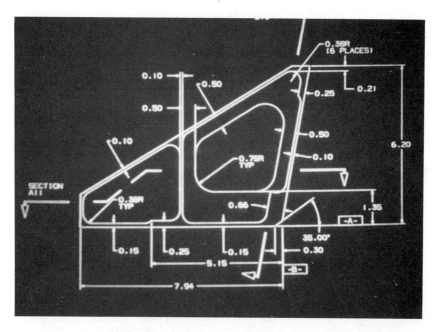

Figure 10.15   Output of front view only. (Courtesy of Auto-Trol Corp.)

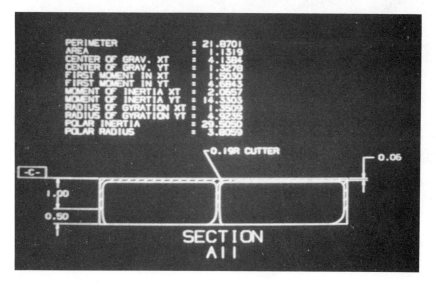

Figure 10.16   Output of Section AII only. (Courtesy of Auto-Trol Corp.)

    The interaction continues in Figure 10.17, which is an output of the
part programmer's working session. Here the notes from Figure 10.16
are used to simulate the production process. Using this process, the
machine part can be produced as shown in Figure 10.18. The difference
in meaning of the terms computer-aided manufacturing, design of machine
elements, and kinetics for machine design is a difference in viewpoint.
It depends on which part of the conversation (interaction) is being considered. In computer-aided kinetics, we use the term mechanism when
dealing with the relative positions and the connections of machine parts.
Also, their relative velocities and accelerations are displayed for
analysis without regard to the shape of the parts (Chapters 1 through 5).
In the computer-aided design of machine elements, a mechanism is a
machine part and is displayed in outline form (Chapters 6 through 10).
Computer-aided manufacturing is not discussed in this chapter other than
to suggest that the sequence is display-analysis-documentation-translation-
production.

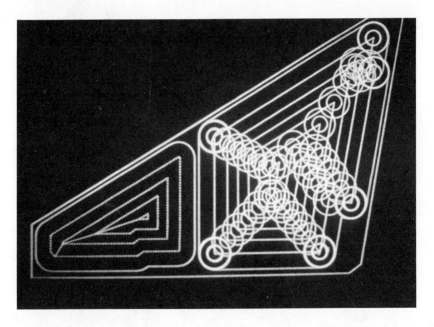

Figure 10.17   Part-programmer working session. (Courtesy of
Auto-Trol Corp.)

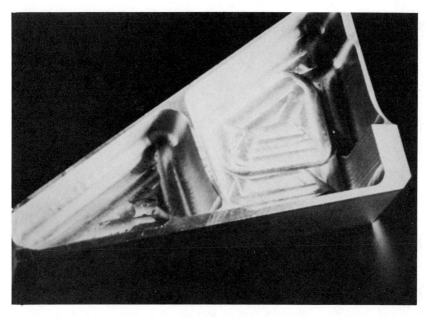

Figure 10.18   Finished machine part. (Courtesy of Auto-Trol Corp.)

AGGREGATE COMPUTER-AIDED COMBINATIONS

Aggregate combinations are where computer displays can be applied to
assemblages of pieces in a mechanism, where the motion of the follower
is the resultant of the motions given by more than one driver. The
number of drivers transmitting motion is two or three. Each driver
determines the motion of one point on the follower, and since the motion
of three points in a body fixes its motion; there can never be more than
three drivers.

Computer-displayed aggregate combinations of drivers and followers
are used to represent very rapid or slow movements and complex paths.
Figure 10.19 illustrates a complex path for a N/C aggregate motion. In
any situation where one driver cannot be used to transmit motion over the
entire motion cycle, aggregate combinations are used. Figure 10.20 is
an example of high-speed aggregate motion displayed on the DVST portion
of the CAMD console.

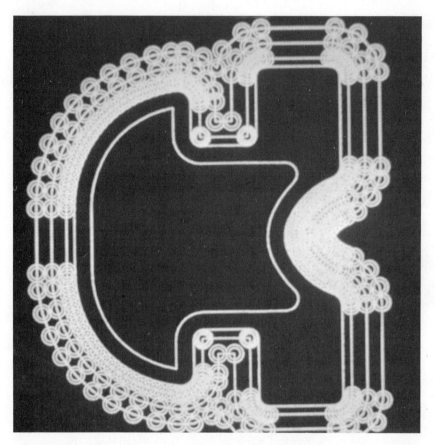

Figure 10.19   Complex-path aggregate motion. (Courtesy of
Auto-Trol Corp.)

Aggregate Motion

Combination motion displays are typical of designs that are to be produced by computer-aided manufacturing means. A typical N/C machine
will have three, four, or even five axis of motion available for the
designer to select. Only three axis of motion are displayed at any
given time. Typically, two types of displays are used:

  1. An outline of the path traveled by the cutter, in this case the
      follower, because the part is driven through space to produce
      a final shape. Figure 10.21 is typical of this method.
  2. An outline of the part traveled during motion, in this case the
      driver. Figure 10.22 is typical of this second display method.

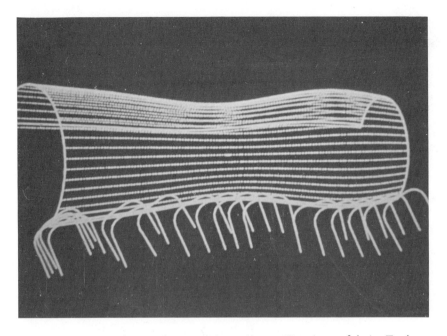

Figure 10.20   High-speed aggregate motion. (Courtesy of Auto-Trol Corp.)

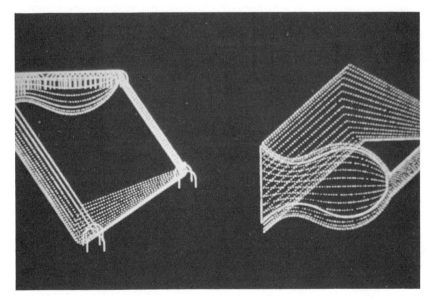

Figure 10.21   Outline of the path traveled by the cutter. (Courtesy of Auto-Trol Corp.)

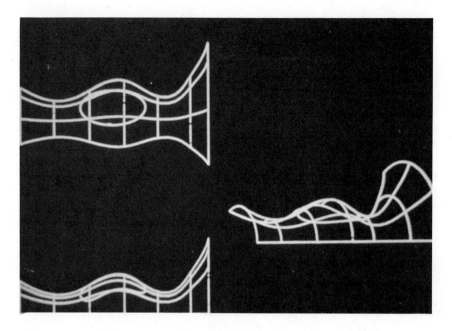

Figure 10.22 Outline of the part traveled during motion. (Courtesy of Auto-Trol Corp.)

Simple Machines

The machine designer employs the concept of aggregate motion together with proven "simple machine" theory to solve more complex problems. For example, Figure 10.23 employs the inclined plane and wedge together with aggregate motion display. Pulley combinations together with aggregate motion displays for hoisting problems are common. Figure 10.24 uses screw, lever, and wheel concepts together with aggregate motion. Finally, Figure 10.25 illustrates all the concepts in a document tation display.

CHAPTER AND BOOK SUMMARY

This chapter has provided a summary of the first nine chapters and a graphic definition of computer-aided design (CAD). CAD is an extremely large field which because of its impact on manufacturing, is one of the revolutionary concepts of the twentieth century. CAD presents not only engineering problems, but also management and production ones, whose effects will, in one way or another, eventually affect all of American

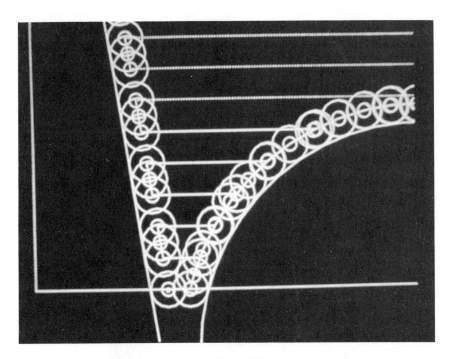

Figure 10.23 Inclined plane and wedge with aggregate motion.
(Courtesy of Auto-Trol Corp.)

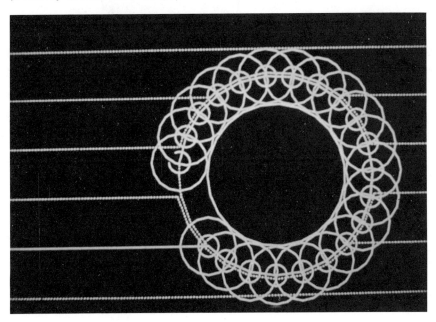

Figure 10.24 Display of screw, lever, and wheel concepts with
aggregate motion. (Courtesy of Auto-Trol Corp.)

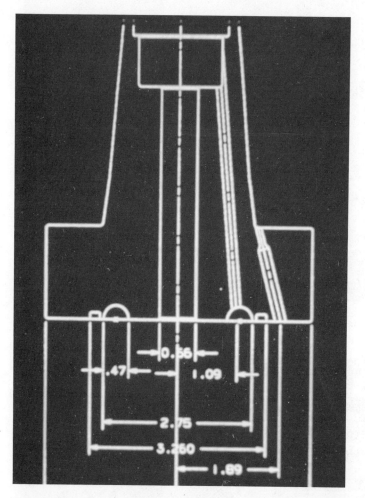

Figure 10.25 Documentation display of simple machine theory.
(Courtesy of Auto-Trol Corp.)

society. A subset of CAD, computer-aided graphics, plays an important
role in this field, for it is from the computer-aided graphics database
that the initial information is obtained for the designer and for the
eventual production of a machine part by a computer-controlled machine
tool.

    This book was written to acquaint the reader with the computer-aided
design of machine elements in both kinematic and part-outline form, and
with the relationships among computer-aided graphics, design, and manufacturing. The reader was not made acquainted with computer-aided

graphics; this is presented in detail in another book published by Marcel Dekker, Computer-Aided Graphics and Design. The reader will gain little from the information in this text if she or he was not familiar with the concepts of computer graphics, machine design, and computer programming before attempting any of the many examples herein that show this relationship. The selection of material in the text was based on the premise that the reader of the first five chapters has had a course in or experience with kinematics; while the second five chapters required an exposure to the design of machine elements. Therefore, many basic design situations were included without lengthy explanations or manual solutions. This kept the emphasis on CAD.

Methods for the use of existing computer-aided kinetics programs were given as well as the procedure for writing new ones. The discussion was, however, on use rather than creation. It is my belief that this delimitation is necessary in a CAD course because CAD requires more ingenuity, inventiveness, imagination, and patience than does display graphics. The ability to create new machine designs will be developed after the reader has experience using the equipment described in this book.

For readers who wish to delve further into certain concepts, a complete bibliography is included at the end of each of the first nine chapters.

Tests

TEST 1. Introduction to Computer-Aided Design

Name_____Seat No._____Section____Grade_____

_____ 1. A computer-generated design requires the answers to:
(A) what machine elements should be used, (B) system
components used, (C) material selection, (D) all of these.

_____ 2. Computer-aided machine design involves two sciences:
(A) physics and chemistry, (B) kinematics and mechanism
design, (C) statics and dynamics, (D) none of these.

_____ 3. In order to display machine elements by computer assistance
the designer should be familiar with the techniques of:
(A) computer graphics, (B) color processors, (C) interactive programming, (D) none of these.

_____ 4. As shown in the typical diagram of CAMD display configura-
tion, the output can be either a CRT presentation or:
(A) a disk record, (B) a plotter drawing, (C) hard copy,
(D) a microfilm image.

_____ 5. A designer may place a point on a member by: (A) CALL
DOT, (B) FA(X,Y), (C) THETA 1=THETA+6, (D) CALL
BETA.

_____ 6. The properties displayed from a computer file in table form
will vary with: (A) treatment during manufacture, (B) size
of structural member, (C) ambient conditions, (D) all of
these.

_____ 7. A structure in CAMD is a fabrication of elements capable of
transmitting forces or: (A) carrying loads, (B) producing
motion, (C) fabrication of floating fasteners, (D) none of
these.

245

_____ 8. A mechanism in CAMD is a combination of rigid bodies so
            arranged that the motion of one: (A) cancels the motion of
            another, (B) produces motion, (C) simulates motion,
            (D) all of these.

_____ 9. For the purpose of computer display the actions of natural
            forces for attraction or repulsion are: (A) written as sub-
            routines, (B) included in the animation program, (C) dis-
            played by a call erase inside a "do loop," (D) not programmed.

_____10. Write a short FORTRAN program to display the moving pairs
           shown.

_____11. For a computer-aided machine element to be displayed as a
           moving body in contact with another body in a definite path, a
           technique is used known as: (A) display pairs, (B) DVST,
           (C) plotter output, (D) COM.

_____12. When subprograms are written to handle all standard elements,
           this is known as: (A) host storage, (B) real-time graphic dis-
           plays, (C) template graphics, (D) interactive mode.

_____13. If one display element not only forms the envelope of the other
           but also encloses it: (A) a lower case exits, (B) a higher case
           is used, (C) a linear translation is displayed, (D) all of these.

_____14. The relative motion of machine elements can be studied where
           both elements move (floating) or one element is usually:
           (A) a fixed piece, (B) an inversion of pairs, (C) exchanged
           with the opposite element of the pair, (D) all of these.

_____15. The computer display of bearing can be used for: (A) straight translations, (B) rotation or turning, (C) helical motion, (D) all of these.

_____16. Display of pulleys that turn freely on cylindrical shafts and at the same time have no longitudinal motion requires: (A) CALL COLLAR, (B) CALL KEY, (C) couplings, (D) none of these.

_____17. A "floating" fastener for a pulley and a shaft may be a key and keyway; this is called: (A) PERT, (B) KEY, (C) BEGIN, (D) feather and groove.

_____18. The arms of a lever may have any angle between them; when the angle is less than 90^0 it is stored as a: (A) bell, (B) toe and wiper, (C) crank, (D) rocker.

_____19. Linkages that are kimematic chains must be studied at a terminal that is capable of: (A) call erase, (B) part animation, (C) multicolors, (D) part simulation.

_____20. Whenever linkages with slot mechanism are used in the computer analysis of machine elements, this is known as a (A) bell-crank, (B) slider-crank, (C) lever-crank, (D) slotted four-bar.

TEST 2. Computer Display of Machine Motion

Name_____Seat No._____Section_____Grade_____

_____ 1. The purpose of computer displays for dynamic illustration of
machine elements is to improve: (A) geometric edges,
(B) the CAD of manufactured products, (C) multisurface
descriptions, (D) machine designers.

_____ 2. A CAMD-DPIDS arrangement provides: (A) high speeds,
(B) reliability, (C) simplicity and economy, (D) all of these.

_____ 3. Part animation implies part motion, which is: (A) a change
of position, (B) a relative term within CAMD, (C) all of
these, (D) none of these.

_____ 4. The display program for motion may be relative or: (A) absolute, (B) fixed in space, (C) curvilinear, (D) none of these.

_____ 5. The motion of a member in CAMD is determined by:
(A) FORTRAN, (B) line or plane, (C) a rectilinear subroutine, (D) three points not in a straight line.

_____ 6. The sense of the display motion is: (A) + or -, (B) along the
line constituting its path, (C) always tangent to the display
curve, (D) all of these.

_____ 7. When a point continues to move indefinitely in a given path in
the same sense, its motion is displayed as: (A) continuous,
(B) contiguous, (C) curvilinear, (D) none of these.

_____ 8. In reciprocating machinery displays, motion consists of:
(A) displacement, (B) velocity, (C) acceleration, (D) all
of these.

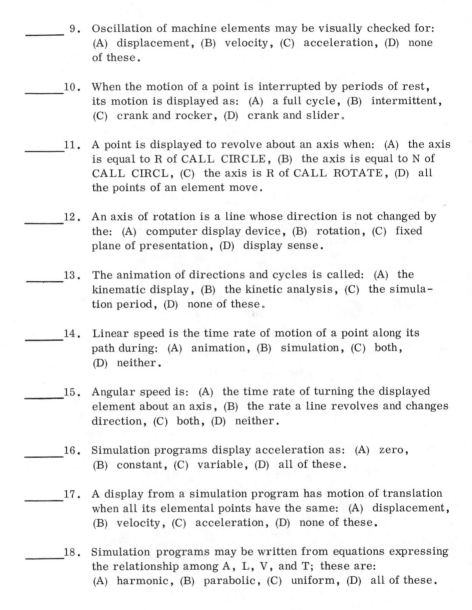

_____ 9. Oscillation of machine elements may be visually checked for:
(A) displacement, (B) velocity, (C) acceleration, (D) none
of these.

_____10. When the motion of a point is interrupted by periods of rest,
its motion is displayed as: (A) a full cycle, (B) intermittent,
(C) crank and rocker, (D) crank and slider.

_____11. A point is displayed to revolve about an axis when: (A) the axis
is equal to R of CALL CIRCLE, (B) the axis is equal to N of
CALL CIRCL, (C) the axis is R of CALL ROTATE, (D) all
the points of an element move.

_____12. An axis of rotation is a line whose direction is not changed by
the: (A) computer display device, (B) rotation, (C) fixed
plane of presentation, (D) display sense.

_____13. The animation of directions and cycles is called: (A) the
kinematic display, (B) the kinetic analysis, (C) the simulation period, (D) none of these.

_____14. Linear speed is the time rate of motion of a point along its
path during: (A) animation, (B) simulation, (C) both,
(D) neither.

_____15. Angular speed is: (A) the time rate of turning the displayed
element about an axis, (B) the rate a line revolves and changes
direction, (C) both, (D) neither.

_____16. Simulation programs display acceleration as: (A) zero,
(B) constant, (C) variable, (D) all of these.

_____17. A display from a simulation program has motion of translation
when all its elemental points have the same: (A) displacement,
(B) velocity, (C) acceleration, (D) none of these.

_____18. Simulation programs may be written from equations expressing
the relationship among A, L, V, and T; these are:
(A) harmonic, (B) parabolic, (C) uniform, (D) all of these.

_____19. Differentiation or integration may be used instead of direct equation solution for simulation, but these cause: (A) display flicker, (B) display jerk, (C) data redundancy, (D) all of these.

_____20. Write a short animation program to display the object shown so that it will be in translation from 0,250 to 1023,250 during 10 display periods.

TEST 3. Computer-Aided Velocity Analysis for CRT Displays

Name_____Seat No._____Section_____Grade_____

_____ 1. Velocities is CAMD may be determined analytically or by:
(A) software, (B) hardware, (C) firmware, (D) graphics.

_____ 2. Velocities are displayed by: (A) resolution and composition,
(B) instantaneous axis, (C) centros, (D) all of these.

_____ 3. The smallest graphical entity in a velocity display is a:
(A) point, (B) line, (C) plane, (D) vector.

_____ 4. Complete vector notation also includes a specification for:
(A) magnitude, (B) direction, (C) sense, (D) all of these.

_____ 5. The display vectors added together to obtain the resultant
are called: (A) components, (B) elements, (C) entity,
(D) none of these.

_____ 6. Display scales are expressed as: (A) proportional sizes,
(B) metric units, (C) English units, (D) all of these.

_____ 7. S(kv) and S(ka) are: (A) velocity scales, (B) subroutines,
(C) animations, (D) simulations.

_____ 8. Dividing a single vector into two components is called:
(A) resolution, (B) combination, (C) composition,
(D) none of these.

_____ 9. The display terms "along" and "perpendicular" are:
(A) resolved components, (B) scalar quantities, (C) vector
composition, (D) none of these.

253

_____10. Each member of a machine is either rotating about a fixed
           axis or: (A) moving axis, (B) instant center, (C) station-
           ary, (D) none of these.

_____11. A centro may be displayed as: (A) a point common to two
           elements, (B) a point about which another element turns,
           (C) a point about which another element tends to turn,
           (D) all of these.

_____12. Centros are displayed on a DVST by: (A) a point-sorting
           routine, (B) three elements having plane motion, (C) three
           elements sharing the same straight line, (D) all of these.

_____13. The method of centros affords an excellent computer display
           for: (A) angular velocity, (B) linear velocity, (C) either
           angular or linear velocity, (D) none of these.

_____14. The sense of rotation for a centro is displayed by: (A) nota-
           tion, (B) sorting, (C) matching, (D) all of these.

_____15. An exchange status command initiates the sequence that
           allows the display controller to access the display list for:
           (A) resolution and composition, (B) instantaneous axis,
           (C) centros, (D) all of these.

TEST 4. Interactive Acceleration Analysis

Name_____Seat No._____Section_____Grade_____

_____ 1. Interactive computer display techniques have made the study
           and analysis of acceleration possible for: (A) moving
           machinery, (B) inertia of elements, (C) high magnitudes,
           (D) none of these.

_____ 2. This new technique for analyzing the accelerations of points
           in the links of a design should be the prerequiste to:
           (A) inertial force analysis, (B) the kinetic cycle,
           (C) visual studies, (D) all of these.

_____ 3. The rate of change in the display velocity in magnitude is:
           (A) the tangential component, (B) the resultant acceleration,
           (C) the tangential acceleration, (D) all of these.

_____ 4. Velocity, polygons can now be constructed by the: (A) 4010
           CRT, (B) graphics tablet, (C) TSO terminal, (D) TTY.

_____ 5. The magnitude of normal acceleration is related to the
           angular velocity of the scalar display and: (A) the distance
           between points, (B) the tangential acceleration about the
           second point, (C) the perpendicular magnitude, (D) none
           of these.

_____ 6. Absolute resultant accelerations originate: (A) at the pole
           point, (B) from the display program, (C) by the designer,
           (D) at the termini of the absolute accelerations.

_____ 7. An acceleration polygon is displayed as: (A) plotter output,
           (B) TSO output, (C) 4010 output, (D) 618 output.

_____ 8.   Klein's construction method is ideal for displaying:
            (A) velocity polygons, (B) slider-cranks, (C) crank-and-
            rocker mechanisms, (D) none of these.

_____ 9.   Display by the Coriolis method is ideal for: (A) points on a
            rolling body, (B) acceleration polygons, (C) crank-and-
            rocker mechanism, (D) all of these.

_____10.   Stated in FORTRAN form, Coriolis's law is: (A) CL = DP/
            (u*w)*COS(THETA), (B) AP = ALP+ALFA+(2.*U*W),
            (C) RESULT = V*(P(I)*FA)/DP, (D) none of these.

Discuss the procedure for the construction of a velocity polygon using a
graphics tablet (use back of sheet if needed).

TEST 5. Computer-Aided Linkage Design

Name_____Seat No._____Section_____Grade_____

_____ 1. When a machine part is displayed as a part of a mechanism, it is called: (A) a link, (B) an element, (C) an entity, (D) none of these.

_____ 2. The subroutine LINK uses two CRT entities PLOT and: (A) DOT, (B) SPOT, (C) CIRCLE, (D) LINE.

_____ 3. The subroutine FA uses two CRT entities CIRCLE and: (A) DOT, (B) SPOT, (C) CIRCLE, (D) LINE.

_____ 4. A combination of CALL FA, PLOT, LINK, PLOT, and FA will display: (A) a crank, (B) a coupler, (C) a follower, (D) all of these.

_____ 5. The subroutine PLOT converts display units to: (A) inches, (B) SI units, (C) meters, (D) none of these.

_____ 6. During the analysis of a four-bar linkage, it is possible: (A) to change the lengths of the links, (B) to invert the pairs or enlarge them, (C) to use sliding pairs, (D) all of these.

_____ 7. A position in the cycle of motion of the driven crank where a straight line exists with the connecting rod is known as a: (A) dead point, (B) high point, (C) slow spot, (D) line hold.

_____ 8. If a series of linkage positions are not erased but are plotted superimposed on one another, and the successive positions are then connected by CALL SMOOT, the locus is called the: (A) spline, (B) FLINE subroutine, (C) IC, (D) centrode.

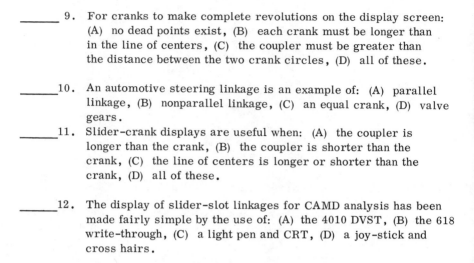

_____ 9. For cranks to make complete revolutions on the display screen:
(A) no dead points exist, (B) each crank must be longer than
in the line of centers, (C) the coupler must be greater than
the distance between the two crank circles, (D) all of these.

_____10. An automotive steering linkage is an example of: (A) parallel
linkage, (B) nonparallel linkage, (C) an equal crank, (D) valve
gears.

_____11. Slider-crank displays are useful when: (A) the coupler is
longer than the crank, (B) the coupler is shorter than the
crank, (C) the line of centers is longer or shorter than the
crank, (D) all of these.

_____12. The display of slider-slot linkages for CAMD analysis has been
made fairly simple by the use of: (A) the 4010 DVST, (B) the 618
write-through, (C) a light pen and CRT, (D) a joy-stick and
cross hairs.

Write a short FORTRAN program to illustrate fixed link motion along the
X axis in a positive direction (use back of sheet if needed).

TEST 6. Computer-Generated Transmission Paths

Name_____Seat No._____Section_____Grade_____

_____ 1. The term computer modeling means that machine elements are
described by: (A) shape, (B) size, (C) turning axis,
(D) all of these.

_____ 2. The display rate, NBAUD, may be varied from 10 to:
(A) 240, (B) 20, (C) 110, (D) 2400.

_____ 3. Crank and slot drives produce nonuniform circular motion in
which the velocity can be manipulated by: (A) computer subroutine, (B) a display program, (C) disk storage, (D) none
of these.

_____ 4. A display program can be written so that the driven slot link
will present a complete revolution with output velocity dependent on: (A) CD, (B) CR, (C) RZ, (D) all of these.

_____ 5. Where BETA is changed from 0° outside the DO LOOP to an
incremental amount inside the LOOP, the result is: (A)
counterclockwise rotation, (B) multiple rotations,
(C) clockwise rotation, (D) all of these.

_____ 6. The display point where the resultant vectors are located is
called the: (A) normal, (B) pitch point, (C) pitch line,
(D) pressure angle.

_____ 7. A technique called "subroutine nesting" requires: (A) additional CPU time, (B) static displays, (C) CALL ROTATE,
(D) no programming knowledge.

_____ 8. The angle the normal line makes with a perpendicular to the line of centers through the pitch point is known as the: (A) common resultant, (B) line of centers, (C) pressure, (D) none of these.

_____ 9. If one element of a pair is in contact with another and the motion is not sliding, it is displayed in: (A) pure rolling contact, (B) common velocity, (C) matched sets, (D) none of these.

_____10. The conditions for displaying rolling contact are: (A) the points of contact must be on the line of centers, (B) the lengths of contacting surfaces must be equal, (C) the display planes perpendicular to the rolling contact must be equal, (D) all of these.

_____11. The display program divides the curve representing the rolling contact into parts so small that the length of the arc is approximately equal to the length of its: (A) chord, (B) smoot, (C) plot, (D) line.

_____12. Write a subroutine for rotating a driver or follower in a clockwise manner (use back of sheet if needed).

TEST 7. Gears and Cams

Name _____ Seat No. _____ Section _____ Grade _____

_____ 1. Gear displays are classified as: (A) parallel, (B) external
(C) internal, (D) spur.

_____ 2. The theoretical point of contact between two gears as they
mesh is displayed as: (A) pitch diameter, (B) pitch circle,
(C) outside diameter, (D) diametral pitch.

_____ 3. The space or clearance displayed between the gear teeth is
known as: (A) addendum, (B) dedendum, (C) working depth,
(D) backlash.

_____ 4. Gear teeth are displayed with two pressure angles, $14\text{-}1/2^{\circ}$
and: (A) $22\text{-}1/2^{\circ}$, (B) 25°, (C) 20°, (D) 12°.

_____ 5. Spur gears can be displayed as circular gears or: (A) straight
lines, (B) hypoidal, (C) miter, (D) herringbone.

_____ 6. The most common form of gear display used is the: (A)
cycloidal, (B) hypoidal, (C) miter, (D) involute.

_____ 7. A cam and its follower form an application of transmitting
motion by: (A) direct sliding contact, (B) pure rolling contact,
(C) friction-free methods, (D) all of these.

_____ 8. Cams may be displayed as: (A) plates, (B) cylinders, (C)
triangular, (D) any combination of plate, cylinder, or
triangle.

_____ 9. The displacement diagram represents the travel of the:
(A) follower, (B) cam during one cycle, (C) speed of the
follower, (D) none of these.

_____10.   In CAMD the movements of parts may be timed by the use of:
(A) cam combinations, (B) gears, (C) displacement diagrams,
(D) computer programs.

TEST 8. Computer-Aided Design of Flexible Connectors

Name _____ Seat No. _____ Section _____ Grade ._____

_____ 1. The CAMD user may select flexible connectors for pulleys from: (A) V-belt drives, (B) flat belts, (C) timing belts, (D) all of these.

_____ 2. The smallest sheave with the highest rate of animation is the: (A) driver, (B) driven, (C) idler, (D) none of these.

_____ 3. The largest sheave with the lowest rate of animation is the: (A) driver, (B) driven, (C) idler, (D) none of these.

_____ 4. The length of the belt at the neutral axis of the display is called: (A) belt pitch length, (B) pitch diameter, (C) center distance, (D) speed ratio.

_____ 5. Combination sheaves are frequently used in CAMD having both ABCDE and: (A) high capacity, (B) standard duty, (C) L series, (D) M series.

_____ 6. Variable speed sheaves are displayed in two general types: (A) manual and spring-loaded. (B) manual and semiauto-matic, (C) manual-adjust and automatic, (D) all of these.

_____ 7. An interactive program will provide the user with: (A) single pulley, parallel axes, (B) open and crossed belt combinations, (C) speed cones and guide pulleys, (D) all of these.

_____ 8. A flexible connector should accomplish a specific objective while performing its work: (A) dampening vibration, (B) absorbing torque, (C) insulating from electrical shock, (D) all of these.

_____9.   Chain drives are not displayed with friction to aid in:
(A) ease of display, (B) transmitting motion, (C) variable
speed, (D) high capacity.

_____10.   The distance in display units from the center of one connec-
ting pin to the next is known as: (A) chain pitch, (B) center
distance, (C) chain rating, (D) pitch diameter.

TEST 9. Computer-Matched Machine Elements

Name _____ Seat No. _____ Section _____ Grade _____

_____ 1. Machine elements designed in the X, Y, and Z axis are
            written in: (A) modular form, (B) matched pairs, (C)
            random order, (D) all of these.

_____ 2. Some 30 or 35 variables must be satisfied to finalize the
            train and: (A) complete the match, (B) gear selection, (C)
            display plots, (D) none of these.

_____ 3. A train is a series of rolling: (A) cylinders or cones, (B)
            gears, (C) pulleys, (D) all of these.

_____ 4. The smallest train is a: (A) display pair, (B) pulley and
            block, (C) rack and pinion, (D) all of these.

_____ 5. The machine designer must match elements in: (A) on-line
            fashion, (B) during real-time, (C) interactive mode, (D) all
            of these.

_____ 6. The use of a graphics tablet coupled with fast response time
            is used for: (A) on-line problem matching, (B) real-time
            selections, (C) interactive selection processes, (D) all of
            these.

_____ 7. The train designer enters the data with a press of the
            graphics tablet pen from the: (A) top menu library, (B) CRT
            face, (C) DVST memory layer, (D) all of these.

_____ 8. A gear located between a driver and driven gear is called:
            (A) an idler, (B) an interfacer, (C) a pinion, (D) none of
            these.

265

_____ 9.   The train value is the reciprocal of the: (A) reverted gear,
             (B) speed ratio, (C) absolute speed difference in first and last
             gear, (D) none of these.

_____10.   The different combinations of speed gear trains are: (A) first
             and last display elements with axes coinciding, (B) driving and
             driven shafts parallel, (C) intermediate parallel shafts, (D)
             all of these.

TEST 10. Computer-Aided Combination of Machine Elements in the Design of Working Machinery

Name _____ Seat No. _____ Section _____ Grade _____

_____ 1. In an industrial setting, a user finds CAD being used for: (A) new products, (B) machines, (C) tools and apparatus, (D) all of these.

_____ 2. The CAD user does: (A) tests, (B) investigations, (C) modeling, (D) all of these.

_____ 3. The purpose of simulation displays for the dynamic illustration of machine elements is to: (A) improve product design, (B) change production methods, (C) slow down the trend toward automation, (D) none of these.

_____ 4. CAMD involves those processes that require a description of: (A) FORTRAN statements, (B) geometric edges, (C) wireforms, (D) primitives.

_____ 5. Sometimes in CAMD surfaces are used to describe: (A) manufacturing methods, (B) 3-D objects, (C) wireforms, (D) primitives.

_____ 6. A digital computer can be used to display tne design, study its movement, and describe its: (A) manufacture by N/C, (B) kinematic actions, (C) kinetic values, (D) none of these.

_____ 7. The part programmer's commands can be transferred by the CAMD software to a suitable: (A) N/C format, (B) CAM instruction code, (C) CNC code, (D) DNC format.

_____ 8.   The intermixed display of refresh and direct-view storage
           tube is known as the: (A) model 618, (B) CRT, (C) DVST,
           (D) 3277.

_____ 9.   In computer-aided kinetics, we use the term mechanism to
           mean: (A) machine, (B) element, (C) parts, (D) all of these.

_____10.   In the computer-aided design of machine elements, a mech-
           anism means: (A) machine, (B) element, (C) part, (D) all of
           these.

Index

Absolute:
  motion, 96
  speed, 97
Acceleration:
  analytic, 75, 224
  angular, 5, 35
  Coriolis', 83
  linear, 5, 34
  normal, 35, 72, 85
  polygons, 79
  relative, 75, 77
  scale, 78
  tangential, 35, 72
  uniform, 34
  vector display, 46, 78, 130
Actions, 15
Addendum, 144
Aggregate computer combinations, 235
Alias, 40
Alloc, 40
Angular:
  acceleration, 5, 35
  displacement, 5
  speed, 33, 98, 132
  velocity, 5
Animation, 16, 25, 39, 227
Annular:
  gear, 156
  pinion, 157
Attrib, 40
Automobile:
  steering mechanism, 104, 105

[Automobile]
  transmission, 217
Axis:
  fixed, 32
  instantaneous, 64
  rotation, 32
  revolution, 32

Base circle, 144
Bearings, 11
Bell crank, 14
Belts:
  design of, 182
  drives, 175
  flat, 175
  length, 176, 181, 182
  nonparallel shafts, 181
  open and closed, 180
  pitch surface, 179
  speed ratio, 177
  timing, 175
  V, 175, 178
Bevel gear, 157
Body:
  rigid, 8
  rolling, 64, 83, 137
Bow's notation, 47

CAD (computer-aided design), 1, 3, 238

CAM (computer-aided manufacturing),
     4, 234
CAMD (computer-aided machine
     design), 1, 4, 229
Cams:
  cylindrical, 157, 159
  diagrams, 159
  eccentric, 158
  flat faced, 158
  motion of follower, 158, 160
  plate, 158, 160
Cedit, 40
Centro:
  location, 65, 66
  notation, 65, 66
  number, 65, 66
  polygon, 67
  velocity of, 65, 67
Centrode, 99
Change, 40
Circular pitch, 144
Code, 40
Collar, 12
Components, 48
Composition, 51
Computer modeling, 121, 141
Cone, 137, 139, 209, 218
Conjugate displays, 131
Constant speed, 26, 34
Copy, 40
Coriolis' law, 83, 84
Cps, 40
Crank displays, 14, 15, 63, 99
Crepro, 40
CRT (cathode ray tube), 2, 11, 45,
     60, 230
Cscript, 40
Curvilinear motion, 30
Cycle (see Kinematic)
Cycloidal, 156
Cylinders, 138
CUCAMD, 18, 22
Cursor, 108

Database, 4, 12
Dead points, 99
Deceleration, 35
Dedendum, 144
Delete, 40
Diametral pitch, 144, 149
Differentiation display, 36
Displacement, 5, 129, 159
Display pairs, 9, 37
Dot, 4
DPIDS (dual processor interactive
     display system), 17
Drag link, 101
Driver and driven, 8, 123, 211
Drum and cable, 192
DVST (direct view storage tube), 9,
     16, 45, 226
Dynamic displays, 69, 225

Edit, 40
Elements, 9, 234
Equal crank, 103
Erase, 99
Exec, 40

FA, 6
Face:
  gear tooth, 144, 147
  width, 144
Feather and groove, 14
Fetch, 41
Fillet, 87
Firmware, 46, 200
Fixed axis, 5, 32
Flank of tooth, 144
Flexible connectors, 92, 175,
     193
Floating link, 61
FORTRAN, 26, 27, 41
Four-bar linkage, 112
Frame, 8

Freeall, 41
Friction gearing, 140, 220

Gear displays, 145, 149, 151, 157
Gear terminology, 144, 157
Graphics tablet, 4, 73, 212
Graphics terminal, 4
GRIP (graphical interactive
    programming), 85
Guide, 11

Hard copy, 4, 12, 129
Harmonic, 36
Helpjcl, 41
Hex, 41
Higher order pairs, 10
Host computer, 4
Hyperboloidal, 144, 157
Hypoid, 157

Ida, 41
Inclined plane, 239
Incomplete pair, 10
Interactive programming, 194
Intergation display, 36
Intermittent (see Motion)
Inversion of pairs, 11
Involute, 153, 172

Joint, 5
Journal, 11
Joy stick, 108

Kennedy's method, 20
Key, 13, 203
Keyboard, 12
Keyway, 13
Kinematic:
    chain, 2

[Kinematic]
    cycle, 33
Klein's method, 20, 81, 83

Label, 41, 92, 95, 205
Lesson, 41
Lever, 14
Line, 87
Linear:
    acceleration, 5, 34
    speed, 33
Link:
    fixed, 94
    floating, 61
    image display, 8, 61
    subroutine, 41, 95
Linkages, 16, 74, 91, 102, 106,
    109
Load, 41
Logoff, 41
Logon, 41
Lower pair, 10

Machine, 2, 7, 91
Mechanism, 2, 7, 91, 234
Memory, 12
Menu, 197
Merge, 41
Motion:
    absolute, 96
    continuous, 27
    curvilinear, 30
    display of, 134, 135
    harmonic, 37
    intermittent, 31
    parabolic, 36
    reciprocating, 27
    relative, 97, 134
    rolling, 134, 137
    uniform, 38

Neck, 11
Nonparallel crank, 102
Nut, 11

Omega, 6, 68, 72, 131
Oscillation, 31, 110
Outside diameter, 144

Pairs, 10
Parallel crank, 102
Part animation, 16, 26, 227
Particle, 8
Passive graphics, 9
Path, 27
Pedestal, 11
Period, 33
Pillow block, 11
Pin joint, 5
Pinion, 155, 157
Pitch:
  circular, 144
  diameter, 144
  point, 128
Pivot, 8, 11
Plot, 93, 96, 196
Point, 5, 83, 87
Polygon, 51, 79, 80, 88
Pressure angle, 144
Print, 41
Process computer, 4
Program, 4, 194
Pulley, 175
Pulley blocks, 189
Purge, 42

Quick return mechanism, 110

Racks, 147, 155
Radian, 35
Rect, 13, 196
Refresh display, 69

Rename, 42
Repro, 42
Resolution and composition, 48, 51,
  62
Resultant, 48
Rings, 12
Rocker, 14, 97
Roller, 132
Rotation, 11, 32, 128

Scalar, 46
Scale, 31, 57, 78
Schedule, 42
Scratch, 42
Screw, 11
Send, 42
Sense, 27
Sheave, 176
Silent chains, 188
Simulation, 33, 36
Slider-crank, 16, 106
Slider-slot, 107, 124
Slist, 42
Software, 4, 69
Speed:
  absolute, 97
  angular, 33, 98, 132
  constant, 26, 34
  linear, 33
  ratios, 121, 146, 177
  uniform, 34
  variable, 34, 183
Spindle, 11
Sprocket and chain, 186, 189
Stepped pulleys, 183
Structure, 7
Submit, 42
Subroutine, 5, 9, 170

Tangential acceleration (see
  Acceleration)
Template storage, 9
Time, 5

Toggle joint, 109
Tooth:
  depth, 144
  face, 144
  flank, 144
  shape, 144
  working depth, 144
Train:
  menu items, 213
  reverted, 215
  value, 214
Translation, 11, 35
Transmission, 8, 183
Turn, 31, 117
Twisted gears, 157

Uniform motion (see Motion)
Uniform speed (see Speed)
Unnum, 42
Us, 42

V-belts, 175
Variable acceleration (see
  Acceleration)
Variable speed (see Speed)
Vector display image, 49, 51, 53,
  69
Vectors:
  acceleration, 46
  addition of, 48
  components, 48

[Vectors]
  direction, 27, 48
  display of, 46
  polygon, 51
  subroutine, 49, 51
  tail-tip, 46
Velocity:
  analysis of, 57
  angular, 5, 73
  image, 51
  linear, 5, 62
  polygon, 74, 80
  relative, 34
  scale, 51

When, 43
Whogot, 43
Whole depth, 144, 232
Working depth, 144, 232
Worm gears, 157
Write-through, 45, 108

X-axis (see Axis)
X-Y plotter, 4, 78

Y-axis (see Axis)

Z-axis (see Axis)